875 - 0475

The Child and Reality

Jean Piaget

Problems of Genetic Psychology

TRANSLATED BY ARNOLD ROSIN

GROSSMAN PUBLISHERS, NEW YORK, 1973

The Child
and Reality

VNYS BF 723 .C5
P528 1973 c.1

CYL235

Problèmes de psychologie génétique

© 1972, by Editions Denoël, Paris-7°

English language translation Copyright © 1973 by The Viking Press, Inc.

First published in 1973 by Grossman Publishers
625 Madison Avenue, New York, N.Y. 10022

Published simultaneously in Canada by
Fitzhenry and Whiteside, Ltd.

SBN 670–21591–0

Library of Congress Catalogue Card Number: 72–93283

Printed in U.S.A.

Contents

The Child and Reality

1 Time and the Intellectual Development of the Child

Child development is a temporal operation par excellence. I will try to offer some data needed to understand this situation.

More specifically, I will focus on two points. The first is the necessary role of time in the life cycle. Any development—psychological as well as biological—supposes duration, and the childhood lasts longer as the species becomes more advanced; the childhood of a kitten and that of a chick are much shorter than that of the human infant since the human infant has much more to learn. This is what I shall try to show in these pages.

The second point is formulated in the questions: Does the life cycle express a basic biological rhythm, an ineluctable law? Does civilization modify this rhythm and to what extent? In other words, is it possible to increase or decrease this temporal development?

To discuss these two points I have in mind only the truly psychological development of the child as opposed to his school development or to his family development; that is, I will above all stress the spontaneous aspect of this development, though I will limit myself to the purely intellectual and cognitive development.

Actually we can distinguish two aspects in the child's intellectual development. On the one hand, we have what can be called the psychosocial aspect, that is, everything the child receives from without and learns in general by family, school, educative transmission. On the other there is the development which can be called spontaneous. For the sake of abbreviation I will call it psychological, the development of the intelligence itself—what the child learns by himself, what none can teach him and he must discover alone; and it is essentially this development which takes time.

Let us immediately take two cases. In a group of objects, for example a bouquet of flowers where you count six primroses and six flowers which are not primroses, you would discover that there are more flowers than primroses, that *the whole is greater than the part*. This seems so obvious that no one would think of teaching it to a child. And yet, as we will see, he needs years to discover such a law.

Another common case concerns transitivity. If a stick

compared to another is equal to it, and if this second stick is equal to a third, is the first—which I will hide under the table—equal to the third? Does *A* equal *C* if *A* equals *B* and *B* equals *C*? Again this is completely obvious to us; no one would imagine teaching this to a child. But he needs, as we will see, almost seven years to discover such logical laws.

Thus what I am going to discuss is this spontaneous aspect of intelligence, and it is the only one I will mention because I am only a psychologist and not an educator; also because from the viewpoint of time, it is precisely this spontaneous development which forms the obvious and necessary condition for the school development.

In our classes at Geneva, it is only around the age of eleven that we begin to teach the notion of proportion. Why not earlier? Obviously, if the child could learn it earlier, the school program would have included the initiation to proportions for those aged nine or even seven. If the child has to wait until he is eleven, the reason is that this notion supposes all kinds of complex operations. A proportion is a relation among relations. To understand a relation of relations, a single relation must first be understood. The whole logic of relations must first be construed, after which this logic of relations must be applied to the logic of numbers. There is a vast set of operations which remain implicit; they are not distinguished at the outset and are hidden beneath this notion of proportion. This example, one of a hundred others possible, shows how psychosocial development is subordinated to spontaneous and psychological development.

I will thus limit myself to this aspect of development and begin at once with a concrete example. It concerns an experiment we did some time ago in Geneva. Offer a child two small balls of clay measuring from three to four centimeters in diameter. The child verifies that they are similar in volume, weight, everything. Ask him to change one of the clay balls into a sausage, to flatten it into a cake, or to divide it into small sections. Then ask him three questions.

First: Has the amount of matter remained the same?

Naturally, you will use a child's language. Ask him if there is the same amount of clay once the ball has been formed into a sausage, or simply if there is more or less clay than before.

Amount of matter, conservation of matter . . . oddly enough, it is usually not until the age of eight that seventy-five percent of children solve this problem. This is only an average. If you conduct the experiment with your own children, you will naturally have a more precocious result, for in relation to the average, your children are certainly ahead. But for the average it is eight years.

Second: Has the weight remained the same?

Offer the child a small scale. If I place the ball of clay on one scale and the sausage on the other, with the understanding that the sausage comes from the ball only by change of form, is the weight going to be the same?

The notion of the conservation of weight is not acquired until about the age of nine or ten, about ten for seventy-five percent of children, that is, at a lag of two years after acquiring the notion of substances.

Third: Has the volume remained the same?

Instead of discussing volume, as the term is difficult, use an indirect method. Immerse the ball in a glass of water; show that the water rises because the ball takes its place. Then ask if the sausage immersed in the glass of water is going to take the same space, that is, make the water rise the same distance.

This problem is not solved till the age of twelve. Thus there is a further lag of two years after solving the problem of the conservation of weight.

Let us quickly see the arguments of those who have no notion of conservation, or of substance, or of weight, or of volume. The argument is always the same. The child will say, "First it was round, then you lengthened the stuff. The moment you made it longer, there is more." He looks at one of the dimensions and forgets the other. The striking thing in this reasoning is that he considers the configuration of the beginning, the configuration of the end, but he fails to consider the transformation itself. He forgets that one thing has been changed into another; he compares the ball seen at the beginning with its state at the end, and he replies, "No, it is longer and therefore there is more of it."

He then discovers that it is the same substance, the same quantity of material. Yet he will say, after the lag of the two years I mentioned and with the same arguments, "It is longer and therefore heavier."

Let us see what the arguments are which lead to the

notion of the conservation. They are always the same and three in number.

The first argument is what I will call the argument of identity. The child will say, "But nothing has been removed, nothing added, so therefore it is the same thing, the same quantity of material." And about the age of eight, he finds it so unusual that one should ask him such an easy question that he smiles, shrugs his shoulders, never thinking that a year earlier he would have given a different answer. He will therefore say, "It's the same thing because you have neither removed nor added anything. But as for the weight, it is longer and therefore heavier." And we have the former argument.

Second argument: reversibility. The child will say, "You have lengthened the matter, you merely have to form it into a ball again and you will see that it is the same thing."

Third argument: compensation. The child will say, "It has been increased in length, that's understood, there is more; but at the same time, it is thinner. The material is more on one side but less on the other. Consequently there is compensation and it's the same thing."

These simple facts at once allow us to make two statements relative to time, by distinguishing in time two fundamental areas: on the one hand, duration, and on the other, order of succession of the events—duration being only the interval between orders of succession.

1. *First of all, time is necessary for duration.* For seventy-five percent of children, one must wait eight years for

the notion of the conservation of substance, and ten for that of weight. Even every adult has not acquired the notion of the conservation of weight. Spencer in his *Treatise on Sociology* tells the story of a woman who preferred to travel with a long piece of luggage rather than a square one because she thought that her dresses when spread out in the long case weighed less than when folded in the square case. As for the concept of volume, one must wait twelve years.

These results occur not only in Geneva—the experiments which we did in Geneva between 1937 and 1940 were resumed in France, Poland, England, the United States, Canada, Iran, and even in Aden, on the banks of the Red Sea. These same stages were found everywhere. But on average no earlier understanding was found with the children of Geneva, even those of a distinguished rank, as we shall soon see. In other words, there is a minimum age for understanding these concepts except, of course, for certain selected social groups, such as children of the talented.

Can one accelerate such development by learning? This is the question that was asked by one of our colleagues, the Norwegian psychologist Jans Smedslund, at our Center of Genetic Epistemology. He tried to hasten acquisition of the notion of conservation of weight through certain instruction—in the American sense of the term— that is, by external reinforcement, for example by reading the result on the scale. But we must first understand that this acquisition of the notion of conservation supposes an entire logic, an entire reasoning based on the transformations themselves, and consequently on the notion of reversibility—this reversibility which the child himself brings to mind when he reaches the notion of conservation. And above all, this notion of conservation supposes transitivity: A state A of the ball being equal to a state B, the state B being equal to a state C, the state A will be equal to the state C. There is a correlation between these various operations. Smedslund began by verifying this correlation and he found a very significant correlation on the subjects studied between the notion of conservation on one side and that of transitivity on the other. He then turned to learning experiences: After each reply, he showed the child the results on a scale which indicated that the weight was the same. After two or three times, the child constantly repeated, "It will always be the same weight, it will again be the same weight," etc.

There will thus be learning of the result. But what is chiefly interesting is that this learning of the result is limited to this result; in other words, when Smedslund turned to the learning of transitivity (which is another matter, since transitivity is part of the logical framework leading to this result), he was unable to obtain learning for this transitivity despite the findings repeated on the scale of $A = C$, $A = B$ and $B = C$. Thus it is one thing to learn a result and another to form an intellectual instrument, a logic required to construct such a result. Such a new reasoning instrument is not formed in a few days. That is what this experiment proves.

2. *Time is equally necessary for order of succession* is the other basic finding which we are going to gain from this example of balls of clay. We have found that the discovery of the notion of the conservation of material precedes that of weight by two years, and that of weight precedes that of volume by two years. This order of succession is found everywhere; it is never reversed; that is, we never find a child who discovers the conservation of weight without having the notion of substance, whereas we always find the opposite.

Why this order of succession? The reason is that if the weight is to be conserved, there must of course be a substratum. This substratum, this substance, will be the material. It is interesting to note that the child begins by the substance, for this substance without weight or volume is not perceptively, empirically noticeable; it is a pure concept but a necessary one in order to continue and arrive at the notion of the conservation of weight and of volume.

Thus the child begins with this empty form which is substance, but he begins here because without it there would be no conservation of weight. As for the conservation of volume, it is a matter of physical and not geometrical volume, including the incompressibility and nondistortability of the substance which, in child logic, will presume its endurability, its mass and, consequently, its weight, since the child distinguishes neither the weight nor the mass.

This order of succession shows that, if a new instrument of logic is to be constructed, there must always be previous logical instruments; that is, the construction of a new notion will always suppose substrata, previous substructures, and this by indefinite regressions, as we will shortly see.

This brings us to the theory of the stages of development. Development is achieved by successive levels and stages. In this development which I am going to describe briefly, we distinguish four important stages.

First, we have a stage, before about age eighteen months, which precedes speech and which we will call that of the sensorimotor intelligence. Secondly, we have a stage which begins with speech and lasts for about seven or eight years. We will call this the period of representation, but it is preoperatory in the sense that I will soon define. Then, between about seven and twelve, we will distinguish a third period which we will call that of concrete operations. And finally, after twelve years, there is the stage of propositional or formal operations.

Thus we distinguish successive stages. Let us note that these stages are precisely characterized by their set order of succession. They are not stages which can be given a constant chronological date. On the contrary, the ages can vary from one society to another, as we will see at the close of this report. But there is a constant order of succession. It is always the same and for the reasons we have just glimpsed; that is, in order to reach a certain stage, previous steps must be taken. The prestructures and previous substructures which make for further advance must be constructed.

Thus we reach a hierarchy of structures which are built in a certain order of integration and which more-over, interestingly enough, appear at senescence to disinte-grate in the reverse order, as the fine work carried out by Dr. Ajuriaguerra and his colleagues seems to show in the present state of their research.

Let us rapidly describe these stages to show why time is necessary, and why so much time is required to reach these notions which are as obvious and simple as those I have used as examples.

Let us begin with the period of sensorimotor intelligence. There is intelligence before speech, but there is no thought before speech. In this respect, let us distinguish intelli-gence and thought. Intelligence for the child is the solu-tion of a new problem, the coordination of the means to reach a certain goal which is not accessible in an immediate manner; whereas thought is interiorized intelligence no longer based on direct action but on a symbolism, the symbolic evocation by speech, by mental pictures, and other means, which makes it possible to represent what the sensorimotor intelligence, on the contrary, is going to grasp directly.

Thus there is an intelligence before thought and speech. Let us take an example. I offer a child a blanket beneath which, without his having seen it, I have slipped a Basque beret. Next I offer the child an object which is new to him, any kind of unfamiliar toy which he wants to grasp, then I hide it under the blanket. He is going to raise the blanket to a certain level in order to find the object, but he fails to see the object, seeing only the beret. He at once raises the beret and finds the object in question. This seems quite simple, but it is a very complex act of intelligence. First, it supposes the permanence of the object. We will soon see that the notion of permanence is not innate but, on the contrary, requires months to be formed. This supposes the localization of the object—which is not given at once, for this localization supposes in turn the organization of the space. This then supposes individual above-below relations, etc. Thus there is an entire construction in this act of intelligence which appears so simple. But an act of intelligence of this kind can be constructed before speech and does not necessarily suppose representation or thought.

Why does this period of sensorimotor intelligence last so long, until about eighteen months?

Another way of posing the same question is: Why is the acquisition of speech so late in respect to the stated mechanisms? Speech has often been reduced to a pure system of conditioning and conditioned reflexes. If such were the case, an infant would acquire speech as early as the end of the first month because the first conditioned reflexes already exist at the beginning of the second month. Why must one wait eighteen months? Our reply is that speech is bound up with thought and thus supposes a system of interiorized actions and even sooner or later a system of operations. We will call operations *interiorized actions,* that is, those no longer carried out materially but from within and symbolically, and actions which can be combined in any manner, especially actions that can be reversed, which are reversible in the sense I just indicated.

As for the actions which form thought, these interiorized actions, one must learn to execute them materially; they first require a whole system of effective and material actions. To think, for example, is to classify, to arrange, to place in correspondence, to collect, to dissociate, etc. But all these operations must be carried out materially in actions, in order to be capable afterward of being constructed in thought. That is why there is such a long sensorimotor period before speech and why speech is late as compared to the development. A long practice of pure action is needed to construct the substructure of later speech.

During this first year, every later substructure is precisely constructed: the notion of object, that of space, that of time, in the form of temporal sequences, the notion of causality, in short, the important notions later to be used by thought and which are developed and used by material action as early as its sensorimotor level.

Let us give two examples. First, there is the notion of the *permanent object*. At first glance, nothing is more simple. The philosopher Meyerson believed that the permanence of the object was given at the very outset of perception, and that there was no way of perceiving an object without believing it to be permanent. Here the infant enlightens us. Take an infant of five or six months after the coordination of vision and prehension, that is, when he can begin to grasp the objects he sees. Offer him an object which interests him, for example a watch. Place it before him on the table and he reaches out to grasp the object.

Screen the object, for example with a piece of cloth. You will see that the infant simply withdraws his hand if the object is not important to him or becomes angry if the object has some special interest for him, for example if it is his feeding bottle. But he does not think of raising the cloth to find the object behind it. And this is not because he does not know how to move a cloth from an object. If you place the cloth on his face, he very well knows how to remove it at once, whereas he does not know how to look behind to find the object. Thus everything happens as though the object, once it has disappeared from the field of perception, were reabsorbed, had lost all existence, had not yet acquired that substantiality which, as we have seen, requires eight years to reach its quantitive characteristic of conservation. The outer world is only a series of moving pictures which appear and disappear, the most interesting of which can reappear when one knows very well how to manage it (for example crying long enough if it is a question of someone whose return is desired). But these are only moving pictures without substantiality or permanence and, above all, without localization.

Second stage: You will see the infant raise the cloth to find the object hidden behind. But the following control shows that the entire notion has not really been acquired. Place the object on the infant's right, then hide it; he is going to look for it. Then remove it from him, pass it slowly before his eyes, and place it at his left. (Here we are

The King's Library

talking of an infant of nine or ten months.) After seeing the object disappear at his left, the infant will at once look for it at his right where he found it the first time. Thus here there is only a semipermanence without localization. The infant is going to look where the action of looking proved successful the first time and independently of the mobility of the object.

Second, there is the situation of *space*. Here again we see that nothing is innate in structures and that everything must be gradually and laboriously constructed. Insofar as space is concerned, the whole sensorimotor development is particularly important and interesting from the viewpoint of the psychology of intelligence. In effect, for the newborn child there is no space that contains objects, since there are no objects (including the body proper which naturally is not conceived of as an object). There is a series of spaces differing one from another and all centered on the body proper. There is the buccal space as described by Stern. For a long time the mouth is the center of the world; Freud has made many remarks on this point. Then there is visual space; but in addition to visual space, there is tactile space and auditive space. And all these spaces are centered on the body proper—the action of looking, of following with the eyes, the action of bringing an object to the mouth, etc.—but they lack coordination with each other. Thus there are egocentric spaces, we might say, not coordinated, and not including the body itself as an element in a container.

Whereas eighteen months later this same child will

have the notion of a general space which encompasses all these individual varieties of spaces, including all objects which have become solid and permanent, with the body itself as an object among others, the displacements coordinating and capable of being deduced and anticipated in relation to the displacements proper.

In other words, for eighteen months, it is no exaggeration to speak of a Copernican revolution (in the Kantian sense of the term). Here there is a complete return, a total decentration in relation to the original egocentric space.

I have spoken enough to show you that eighteen months are quite inadequate to construct all this and that actually this development is singularly accelerated during this first year. This is the period of childhood when acquisitions are most numerous and rapid.

I will move on to discuss the period of the preoperatory representation. At about a year and a half or two years, an important event occurs in the child's intellectual development. It is then that there appears the capacity to represent something with something else, which is known as the symbolical function. One form of symbolical function is speech, a system of social signs as opposed to individual symbols. But simultaneous with this speech, there are other manifestations of the symbolical function. A second form is play which becomes symbolical: representing something by means of an object or of a gesture. Until then play was only a play of motor exercises, whereas after about a year and a half, for example, the child really begins to play.

One of my children passed around a seashell in a box while saying "meow" because a moment earlier he had seen a cat on a wall. In this case the symbol is obvious, since the child had no other word at his disposal. What is new, however, is to represent something with something else.

A third form of symbolism can be a gestural system of symbolics, for example in the postponed imitation.

A fourth form will be the beginning of the mental picture or interiorized imitation.

Thus there exists a set of symbolizers which appear at this level and which make thought possible, thought being, I repeat, a system of interiorized action and leading to those particular actions which we will call *operations,* reversible actions and actions coordinating one to another into a total system about which I will shortly have a few words to say.

A situation is presented here which poses the problem of time in the most acute manner. Why do the logical structures, the reversible operations which we have just characterized, and the notion of conservation which we have mentioned, not appear the moment there is speech and the moment there is symbolical function? Why must we wait eight years to acquire the invariant of substance and more so for the other notions instead of their appearing the moment there is symbolical function, that is, the possibility of thought and not simply material action? For the basic reason that the actions that have allowed for certain results on the ground of material effectivity cannot be interiorized any further in an immediate manner, and that it is a matter of relearning on the level of thought what has already been learned on the level of action. Actually this interiorization is a new structuration; it is not simply a translation but a restructuration with a lag which takes a considerable time.

I will offer an example: It is the group of displacements which, in the sensorimotor organization of space, constitutes a basic final result. What geometers call a group of displacements is, for example, that the child, while circulating in his apartment or in his garden when he learns how to walk, is able to coordinate his coming and going, to return to the point of departure (reversibility) or to make detours to arrive at the same point by different ways (associativity of the group of displacements). In short, he is going to coordinate his displacements in a whole system which allows him to return to the point of departure.

As early as the age of about a year and a half, this group of displacements is acquired on the sensorimotor level. But does this mean that the child can represent for itself, in a mental picture or by drawing or speech, the displacements that he knows how to do materially? Not at all. Because to move about is one thing and to summon to mind the same displacements by representation is something quite different.

Together with my colleague Szeminska, we once did a most interesting experiment on children aged four and five who, at a time when there was less traffic in Geneva, went alone from home to school and returned alone from school to home two or four times a day. We tried to have them represent the path they took between school and home not by means of a drawing, since this would have been too complicated, nor by speech, which would have been even more difficult, but by means of a small construction game. We had a blue ribbon for the Arve, a green box for the Plainpalais Plain, we showed the church at the far end of the plain, the Exhibition Palace, etc., and the child was asked to place the different buildings in regard to his house and school. These children aged four and five knew how to follow their way to school, but they were unable to represent it; they gave it a kind of motory representation. The child said, "I leave the house, I go like that (gesture), then like that (gesture), then I make a turning like that, then I reach school."

But to place the buildings and lay out the way was a completely different matter. It is one thing to get about somehow in a foreign city after just arriving and to know how to get about after a few days, and it is something else to imagine the plan if there is no city plan available. Whether one and the same action is done materially or summoned to mind, it is really not the same action. The development is not linear: A reconstruction is necessary. This explains why there is a whole period which lasts until about the age of seven or eight when what has been acquired on the sensorimotor level cannot be continued but must be elaborated again on the level of representation, before leading to these operations and conversations which we spoke of earlier.

*   *   *

I now come to the level of concrete operations, at an average age of about seven years in our civilizations. But we will see that there are delays or increases due to the action of social life. About the age of seven, a fundamental turning point is noted in a child's development. He becomes capable of a certain logic; he becomes capable of coordinating operations in the sense of reversibility, in the sense of the total system of which I will soon give one or two examples. This period coincides with the beginning of elementary schooling. Here again I believe that the psychological factor is a decisive one. If this level of the concrete operations came earlier, elementary schooling would begin earlier. This is not possible before a certain level of elaboration has been achieved, and I shall now try to give its characteristics.

Let us note at once that the operations of thought, on this level, are not identical to what is our own logic or to what adolescent logic will become. Adolescent logic—and our logic—is essentially a logic of speech. In other words, we are capable—and the adolescent becomes so as early as the age of twelve or fifteen—of reasoning on propositional, verbal statements. We are capable of manipulating propositions, of reasoning by placing ourselves in the viewpoints of others without believing the propositions on which we reason. We are capable of manipulating them in a formal and hypothetico-deductive manner.

As we will see, this logic requires much time to be constructed. Before this logic, a previous stage must be passed, and this is what I will call the period of concrete operations. This previous period is that of a logic which is not based on verbal statements but only on the objects themselves, the manipulable objects. This will be a logic of classifications because objects can be collected all together or in classifications; or else it will be a logic of relations because objects can be combined according to their different relations; or else it will be a logic of numbers because objects can be materially counted by manipulating them. This will thus be a logic of classifications, relations, and numbers, and not yet a logic of propositions. Nevertheless we are dealing with a logic, in the sense that for the first time we are in the presence of operations that can be reversed—for example addition, which is the same operation as subtraction but in a reversed way. It is a logic in the sense that the operations are coordinated, grouped in whole systems which have their laws in terms of totalities. And we must very strongly insist on the necessity of these whole structures for the development of thought.

A number, for example, does not exist in an isolated state. What is given is a series of numbers, that is, an organized system which is the unit plus the unit, and so forth. A logical classification, a concept, does not exist in an isolated state. What is given is the total system which we will call a classification. Likewise a "greater than . . ." relation of comparison does not exist in the isolated state; it is part of a whole structure we will call the seriation, which consists in arranging the elements according to the same relation.

It is these structures which are constructed as early as

the age of seven, and from this moment on the notions of conservation become possible.

Let us take two examples of these whole structures.

1. *Seriation*. Give a child a series of sticks which differ in size and ask him to arrange them in order, from the smallest to the largest. Before the age of seven, he will of course be able to do so but in an empirical manner, that is, with hesitation, which is not a logical operation. However, from the age of seven on, he is capable of a system. He will compare the different elements until he finds the smallest which he places on the table, then he will look for the smallest of those remaining and place it alongside the first, then the smallest of those remaining and he will again place it alongside the second, each element being both larger than those already placed and smaller than those remaining: Here is an element of reversibility.

This modest operation is acquired about the age of seven on the level of lengths. If you express this operation in terms of pure language, it becomes far more complicated. In the Burt intelligence tests, which are so rich in logical operations, there is the following test which I once studied with much interest. It concerns three young girls who differ because of the color of their hair, and the question is which has the darkest hair of the three. Edith is lighter than Suzanne and at the same time darker than Lili. Which is the darkest of the three? A certain reasoning is required which is not immediate, even with an adult, to discover that it is Suzanne and not Lili. The child must wait twelve years until he can solve this problem, because it is given in terms of verbal statements. And yet there is nothing more than the seriation which I just mentioned, but it is a verbal seriation which is something other than the concrete operations which I am describing.

2. *Classification.* This is acqiured only about the age of seven or eight, if you take as criteria of classification the inclusion of a subcategory in a category, that is, the understanding of the fact that a part is smaller than the whole. This can appear extraordinary and yet it is true. Give a child flowers which include six primroses and six other flowers. Ask him, "Are all the primroses flowers?" He'll reply, "Of course; yes." "Are all the flowers primroses?" His reply, "Of course not." "Are there more primroses on this table or more flowers?" The child will look, then reply, "There are more primroses," or he will say, "It's the same thing because there are six on one side and six on the other."

"But you just told me that primroses are flowers. Are there more flowers or more primroses?"

The flowers mean to him what remains after the primroses; this is not the inclusion of the part in the whole but the comparison of a part with the other part.

This is interesting as a sign of concrete operations. Note that with flowers, this problem is solved about the age of eight. But if you take animals, the solution comes later. Ask a child, "Are all animals birds?"

"Certainly not. There are snails, horses. . . ."

"Are all birds animals?"

"Certainly."

"Well, if you look out the window, are there more birds or more animals?"

"I don't know. You'd have to go and count them."

It is impossible therefore to deduct the inclusion of the subcategory in the category simply by the manipulation of the "all" and "some." And this probably because the flowers can be gathered in bouquets. This is an easy concrete operation, whereas to go and make a bouquet of swallows becomes more complicated; it is not manipulable.

I come finally to the formal operations at about the age of twelve and with fourteen to fifteen years of age as equilibrium level.

This concerns a final stage during which the child not only becomes capable of reasoning and of deducting on manipulable objects, like sticks to arrange, numbers of objects to collect, etc., but he also becomes capable of logic and of deductive reasoning on theories and propositions. A new logic, a whole set of specific operations are superimposed on the preceding ones and this can be called the logic of propositions. Actually this supposes two fundamental new characteristics. First there is the *combinatory*. Until now everything was done gradually by a series of interlockings; whereas the combinatory connects any element with any other. Here then is a new characteristic based on a kind of classification of all the classifications or seriation of all the seriations. The logic of propositions will suppose, moreover, the combination in a unique system of the different groupments which until now were based either on reciprocity or on inversion, on the different forms of reversibility (group of the four transformations: inversion, reciprocity, correlativity, identity). Thus we are in the presence of a completion which, in our societies, is not noted until about the age of fourteen or fifteen and which takes such a long time, because, to arrive at this point, the child must go through all kinds of stages, each being necessary to the achievement of the following one.

Until now I have tried to show the necessary role of time in the child's intellectual development. I am now going to speak of the other question which we asked ourselves at the beginning of this study, namely, is it a matter of an ineluctable rhythm, or are there variations possible under the effect of the civilization or under that of the societies in which the child lives?

Two replies can be given: the actual reply and the reply of theoretical interpretation. But unfortunately the actual replies are inseparable from the theoretical interpretation, because a fact is nothing in itself if it is not interpreted and the interpretation here is always delicate.

The acutal state: We of course find accelerations in relation to the ages I have just given. Some individuals are gifted more than others. From time to time there are geniuses. Thus there are accelerations, but are these accelerations the result of a more rapid biological maturation? This is quite possible, for there are very different rhythms in individual growth. Or else are they an effect of education, exercise, etc? Here the raw fact makes it impossible to reply and an interpretation is needed.

Moreover we will find collective accelerations in certain social classes and in certain milieus. But here again, is˘it a matter of a selection of those gifted or of a social action?

As a matter of fact, what we find, especially in the comparative studies made in every kind of country on every kind of result, are surprising delays in relation to the ages which we have given. For example, the Canadian psychologists, who went back to these proofs in detail and in a very standardized manner, found in Montreal practically the same ages as in Geneva. But in giving the same tests in Martinique, they found a four-year lag in the replies given to all our problems. And yet it was a question of children schooled according to the French program of elementary teaching, which goes through the certificate of elementary studies. Nevertheless Martinique children have a four-year lag in the acquisition of the notions of conservation, deduction, seriation. . . .

But what is the question here? Does this delay stem from a factor of maturation, in other words, from a racial factor? This seems scarcely possible because, psychologically, nothing similar has ever been found. Or is it a question of a social factor, that is, of a certain passivity in the adult social milieu? The psychologists whom I am going to mention (A. Pinard, M. Laurendeau, C. Boisclair) would prefer to turn toward this second direction by furnishing in this respect all sorts of indications. One of the children's teachers who was studied had greatly hesitated before choosing his profession between the vocation of teacher and another possible one, that of sorcerer. . . . An adult milieu lacking a dynamic intellectual quality can be the cause of a general delay in child development.

On the other hand, research was done in Iran. The same ages were found in Tehran as here; but among country illiterates, a few hours from this city, a two-and-a-half-year delay was noticed, and this in an almost constant manner. The order of succession remains the same but with lags.

Here therefore is the state of fact: There are variations in the rapidity and duration of the development. How to interpret them? The development, of which I have tried to make a very schematic and very concise list, can be explained by different factors.

I distinguish four.

First factor: heredity, internal maturation. This factor can certainly be retained from every viewpoint, but it is insufficient because it never occurs in the pure or isolated state. If a maturation effect intervenes everywhere, it remains dissociable from the effects of the exercise of learning or of experience. Thus heredity is not a factor which acts alone or which can be isolated psychologically.

Second factor: the physical experience, the action of objects. It again forms an essential factor which cannot be underestimated but which likewise is insufficient. Child logic especially is not drawn from the experience of objects, it is drawn from the actions which effect the objects. This is not the same; that is, the part of the child's activity is fundamental, and here, the experience drawn from the object is insufficient.

Third factor: social transmission, the educative factor in the large sense. Naturally a determining factor in development, it alone is insufficient for the obvious reason that if a transmission is possible between adult and child or between the social milieu and the educated child, the child must assimilate what one is trying to inculcate in him from without. This assimilation is always conditioned by the laws of this partially spontaneous development, a few examples of which I have given.

Let us recall in regard to this the inclusion of the subcategory in the category, the part smaller than the whole. Speech contains many cases in which the inclusion is marked in a completely explicit manner by the words themselves. But this does not enter the child's mind so long as the operation is not constructed on the level of the interiorized actions. For example, I once studied—this was another Burt test—a test in which it was a question of determining the color of a bouquet of flowers, considering the following statement:

A boy says to his sisters, "Some of my flowers are buttercups. (I had even simplified by saying, "Some of my flowers are yellow.") The first sister replied, "Then your bouquet is yellow, it is completely yellow." The second replied, "Part of your flowers are yellow"; and the third replied, "None of your flowers is yellow."

Young French children—this was a study made in Paris—until the ages of nine and ten replied: "The first two are right because they say the same thing. The first said, 'Your whole bouquet is yellow' and the second, 'Some of your flowers are yellow.' That's the same thing; that means that there are some yellow flowers and that they are all yellow." In other words, the genitive partitive, the relation of the part to the whole, was not understood by speech for lack of structuration of the inclusion.

I would like to speak of a fourth factor which I will call the factor of equilibrium. From the moment when there are three factors, there must already be a balance among them; but further, in intellectual development, a fundamental factor intervenes. The fact is that a discovery, a new notion, a statement, etc., must balance with the others. A whole play of regulation and of compensation is required to result in a coherence. I take the word *equilibrium* not in a static sense but in that of a progressive equilibration, the equilibrium being the compensation by reaction of the child to the outer disturbances, a compensation which leads to the operatory reversibility at the end of this development.

Equilibrium appears to me to be the fundamental factor of this development. We then understand both the possibility of acceleration and at the same time the impossibility of an increase going beyond certain limits.

The possibility of acceleration is given in the facts which I previously indicated, but theoretically, if the development is above all a matter of equilibrium, because a balance can regulate itself more or less rapidly according to the child's activity, it is not regulated automatically like a hereditary process which would be subjected from within.

If we compare to the young Greeks of the time when

Socrates, Plato, and Aristotle invented the formal or prop-ositional operations of our Western logic, our young con-temporaries, who have to assimilate not only the logic of proportions but all the knowledge acquired by Descartes, Galileo, Newton, and others, a hypothesis must be made that there is a considerable increase in the course of child-hood until the level of adolescence.

Balance takes time, this we agree, but the equilibra-tion can be more or less rapid. Nevertheless this acceler-ation cannot grow indefinitely, and it is here that I will end. I do not believe that there is even an advantage in at-tempting to increase child development beyond certain limits. Balance takes time and everyone portions this in his own way. Too much increase runs the risk of interrupt-ing the balance. The ideal of education is not to teach the maximum, to maximize the results, but above all to learn to learn, to learn to develop, and to learn to continue to develop after leaving school.

2 Affective Unconscious and Cognitive Unconscious[1]

In this report I will deal with the problems of the unconscious and of the conscious as they are encountered in the study of intelligence, representation, and cognitive functions, because I believe particular questions relative to the cognitive unconscious are parallel to those which, in psychoanalysis, are raised by the functioning of the affective unconscious. My aim is, of course, neither to try to introduce some novelty into the psychoanalytical theories nor to criticize them, because if I am somewhat heretical in my viewpoint, this is not the place to explain. Furthermore, I am convinced that there will be a time when the psychology of the cognitive functions and psychoanalysis will have to blend into a general theory which will improve both by

---

[1] Lecture given in full session at the American Society of Psychoanalysis.

correcting each, and it is this future that is worth preparing by showing presently the relations which can exist between them.

The problem of structures

Affectivity is characterized by its energetic compositions with loads distributed to one object or another (*cathexis*) according to the positive or negative connections. On the contrary, what characterizes the cognitive aspects of behavior is their structure, whether it is a question of schemes of elementary actions, concrete operations of classifications or seriation, etc., or of logic of proportions with their different *foncteurs* (implications, etc.). The result to which affective, therefore energetic, operations, lead is simply conscious; that is, it is expressed by feelings which the subject feels more or less clearly as given data. However, the intimate mechanism of these operations remains unconscious; that is, the subject knows neither the reasons for his feelings, nor their source (their entire association with the subject's past), nor the reason for their more or less strong or weak intensity, nor their eventual ambivalences, etc. It is this intimate and hidden functioning of the energetic compositions that psychoanalysis tries to free, and it is not for me to remind you how complex this affective unconscious is with its wealth of content and intricacy of dynamic networks.

My role is to point out that, in the case of cognitive structures, the situation is remarkably comparable: some (but rather limited) consciousness of the result and almost entire (or initially entire) unconsciousness of the intimate mechanism leading to the result. The result is somewhat conscious in the sense that the subject rather knows what he thinks of an object or of a problem, and he knows his own opinions and beliefs, especially insofar as he manages to formulate them verbally in order to communicate with others or to oppose different opinions. However, that only concerns the result of the intimate functioning of the intelligence, and the intimate functioning remains entirely unknown to the subject until, at very superior levels, thought on this problem of structures becomes possible. Until then, the subject's thought is directed by structures whose existence he ignores and which determine not only what he is capable or incapable of "doing" (hence the extents and limits of his power to solve problems) but also what he "must" do (hence the necessary logical connections which are imposed on his thought). In short, the cognitive structure is the system of connections that the individual can and must use, and is in no way the contents of his conscious thought, since it is he who imposes certain forms rather than others, and this according to successive levels of development whose unconscious source goes back as far as the organic and nervous coordinations.

Thus the cognitive unconscious consists of a set of structures and functionings ignored by the subject except for their results. It was for profound reasons then that Binet [2] once stated this truth as though it were a whim:

---

[2] Alfred Binet (1857–1911), French psychologist, born at Nice. He studied physiological psychology and experimental psychology. His work included the origin of the mental test method.——TRANS.

"Thought is an unconscious activity of the mind." What he wanted to say was that if the self is conscious of the contents of its thought, it knows nothing of the structural and functional reasons which force it to think in this or that manner, in other words, of the intimate mechanism which directs thought.

What I maintain then is by no means particular to infant thought; it is found again not only in every adult thought but also in the development of scientific thought. Thus mathematicians of all times have reasoned by obeying, without knowing it, the laws of certain structures, the most exacting of which is the structure of the group, which is easy to see in the work of Euclid's *Elements,* for example. Mathematicians knew nothing of this, and it was only in the early nineteenth century that Galois [3] "became aware" of the existence of such a structure now recognized by everyone as fundamental. Similarly, Aristotle, in creating logic by concentrating on the manner in which he, as well as his contemporaries, reasoned, "became aware" of some of the simplest structures of the logic of category and of syllogism. What is very interesting, however, is that at the same time he was not aware of a whole set of structures which he himself used and which are the logic of relations: This also was not understood until the nineteenth century with Morgan's work, etc.

It goes without saying that if there is unconsciousness

---

[3] Evariste Galois (1811–1832), French mathematician, born at Bourg-la-Reine, and killed in a senseless duel. He is noted for his group substitutions and theory of functions.——*Trans.*

on the level of scientific thought, one of the goals of which is the study of structures, we will find unconscious again, but far more systematically, in all other forms of thought: in the case of the "natural" thought of one normal adult who does not specialize in sciences, and *a fortiori* in the case of spontaneous and ever-creative thought which characterizes the child at the different levels of his development.

Let us be satisfied with a single example in the child: that of the structure of transitivity. When, still at the age of five or six, the child is shown two sticks A and B, such that A is smaller than B, then shown stick C which is larger than B while A is hidden, the child fails to deduce the relation A is smaller than C since he does not see simultaneously A and C. However, between the ages of six and seven the structure of transitivity is constructed and it is then successfully applied to many different problems not only of a causal order but also of a mathematical or logical one. But the child himself is wholly unaware that he constructed such a structure and believes to have reasoned in the same manner. He knows even less on what this structure (groupment of relations) is based, nor does he know how or why it became necessary for him. In short, he is aware of the results he obtained but in no way aware of the intimate mechanisms which transformed his thought, whose structures as structures remain unconscious. It is to these mechanisms as structures that, functionally, we will give the overall name of *cognitive unconscious*.

*The awareness of one's own action
and the cognitive repression*

Let us now study a few special actions on the child's part,
no longer as submitted to the underlying structures but
rather as *revealed contents* (to use a Freudian term) and
which should all be conscious since they simply form a result of the hidden functioning of the mind and not a part
of this functioning itself. We are going to observe in this
area that, if the awareness is generally easy, there are cases
where awareness is counteracted by an inhibitory mechanism which we could compare to an affective *repression*
(a notion which is among the great discoveries of Freudian
psychoanalysis).

As an example of special action whose awareness is
easy, we can discuss the one which consists of throwing an
object into a box at a given distance: Children, about age
four, that one questions, say that they were placed in front
of the box and if they are put at its side, they turn themselves to remain facing the goal. Let us now see what
occurs in the following experiment (and of course without
having beforehand raised the preceding problem in order
to exclude all influence of eventual suggestion). A child is
given a sling in its simplest form: a ball attached to a string
which is whirled, then aimed at a goal.[4] At first, there is no
goal whatever and the child enjoys whirling the ball at the
end of the string and then letting it go, noting that it flies

---

[4] Research made with A. Fluckiger.

off from his side (and in general even seeing that it flies off in the extension of the rotary direction). Next a box is placed thirty to fifty centimeters away and the child, often as early as five years old, quickly manages to reach the box by whirling the ball from his side (about nine o'clock, if we consider as clock dial the rotation surface, the box itself being placed at noon). Having done so, the child is complimented; he begins again several times and is asked where he has released the ball.

A strange reaction then occurs. The youngest children claim that they released the balls exactly in front of them (about six o'clock) and that the ball left in a straight line, from six o'clock to noon (the diameter of the rotary circle) into the box. Others (children aged seven to eight) claim that they released the ball at noon, that is, facing the box. About the ages of nine to ten, there are often compromises: The ball is released about eleven or ten-thirty, and it is only about the age of eleven or twelve that the child replies at once that the ball left at nine o'clock, that is, tangentially and no longer facing the goal. In other words, the child soon knows how to accomplish a successful action, but years are needed before he becomes aware of this, as if some factor were opposed to this knowledge and retained in the unconscious certain movements or even certain intentional parts of successful behavior.

The factor of inhibition is easy to discover. The child represents his own action as divided into two periods: spinning the ball, then throwing it into the box; whereas without this goal he throws the moving object anywhere.

But, for him, throwing to the goal supposes a perpendicular trajectory to the box, thus a release facing it. When asked to describe his action, he thus reconstructs it logically as a function of this preconceived idea and hence does not wish to see that actually he proceeded differently. Therefore he distorts or even dismisses an observation contrary to the idea he has and which alone seems right to him.

Many other similar cases could be mentioned. When a child manages (alone or by imitation), using one or two fingers, to throw a ping-pong ball on a horizontal plane in such a manner that it returns, he does not want to see that he made it turn around from the very beginning and he thinks he notices that it first goes forward, then itself changes direction.[5] Or again when an infant pushes an object with a stick by touching its side, he does not see that he imprints on it movements both of rotation and of transfer, etc.

To explain these gaps or distortions of awareness, one would be tempted to turn to an apparently very clear reason: The child simply does not "understand" what he has done and limits himself to retaining only what appears intelligible. We believe, however, that this interpretation remains insufficient. It is untrue that the child has understood nothing of his successful actions (tangential movement of the ball launched by the sling, reverse rotation of the ping-pong ball, etc.): He understood the essential but *in action* and not in thought, hence by sensorimotor schemes and not representative ones. In other words, he

---

[5] Research made with A. Papert-Chrystrophider.

"knows" how to launch these projectiles, etc., and he knows this as a function of a certain perceptive-motor learning, and not at all from an innate manner.

However, the problem remains: Why do certain sensorimotor schemes become conscious (by an expression in representative and even verbal concepts), whereas others remain unconscious? The reason is that the latter contradict certain former conscious ideas (for example, that the child must face the box to launch a ball or that a ball will not advance by turning in an opposite direction, etc.) and that the sensorimotor scheme used and the former previous idea are therefore incompatible. In these cases, the scheme of action cannot, of course, be integrated into the system of conscious concepts, and it is therefore eliminated, since these concepts, already conscious and long received, are superior in rank to the scheme.

We therefore find ourselves in a situation quite similar to that of affective repression: When a feeling or an impulse finds itself in contradiction with the feelings or tendencies of a superior rank (emanating from the super-ego, etc.), they are then eliminated, thanks to two kinds of operations: a conscious repression or an unconscious repression. We observe here an analogous mechanism on the cognitive ground, and it is really an unconscious repression that we are considering: Actually the child has not first made a conscious hypothesis in order later to dismiss it: He has, on the contrary, dismissed the awareness of the scheme, that is, repressed the scheme from the field of the conscious before it penetrates there in a conceptualized form (and we will soon see that no other is possible, for even a mental picture refers to a concept).

Moreover this mechanism of cognitive repression is without doubt more general than awareness of action (hence sensorimotor schemes). On the neurological level, for example, Pribram showed that a cortical mechanism of regulation, in contact with several inputs, retains some of them, which become stimuli, and "dismisses" others which then cannot act.

The mechanism of awareness

The common definition gives a wholly insufficient (if not erroneous) idea of awareness by representing it as a kind of illumination which reveals realities until then obscure yet changes nothing whatever (in the manner that a flashlight turned to a corner suddenly makes everything visible yet modifies neither the positions nor the relations of the objects). Awareness is much more than that, since it consists of bringing certain elements from an unconscious lower level to a conscious upper level. These two levels cannot be identical, otherwise there would be no problem and the passage would prove an easy one, which is not the case. Hence awareness consists of a reconstruction on an upper level of what is already organized in another manner on a lower level, and the two problems are that of the function of this reconstruction and that of its structural process.

From the viewpoint of function, Claparède noticed

that awareness occurs when there is a disadaptation, for when a behavior is well adapted and functions easily, there is no reason to analyze its mechanism in a conscious manner. Hence we can quickly walk down a flight of steps without representing to ourselves every leg and foot movement; if we do, we run the risk of compromising this successful action. In one of my studies, A. Papert had children walk on all fours, then asked them to describe the movements made by their hands and feet. The children's first descriptions were not possible (both hands advancing at the same time, then both feet, etc.). Then came a possible but nonutilizable description (both left limbs, then the right ones, etc.); and only two-thirds of the children aged ten to eleven described what they had done properly. Before revealing these results at one of our symposiums, Mme. Papert asked our honorable guests to forsake their dignity and take part in an experiment of walking on all fours. Physicists and psychologists correctly analyzed their movements, but logicians and mathematicians reconstructed them deductively according to the second scheme (possible but nonutilizable).

If a well-adapted action requires no awareness, it is directed by sensorimotor regulations which can then automate themselves. When, on the contrary, an active regulation becomes necessary, which supposes intentional choices between two or several possibilities, there is awareness in function of these needs themselves. This is what the previously mentioned research demonstrates.

As for the structural process, the reconstruction forming the awareness consists of a conceptualization. Actually, the cognitive unconscious does not include such concepts as representation, and the very idea of *unconscious representations* seem to me contradictory even though it is current: The unconscious is equipped with sensorimotor or operatory schemes already organized as structures but expressing what the subject can "do" and not what he thinks. From the affective viewpoint, he is likewise equipped with energetic tendencies and loads, hence affective or characterial schemes, etc.

The conceptualized reconstruction which characterizes awareness can at first be sufficient when it is not inhibited by any contradiction. If not, first it is distorted and wanting, then gradually self-completing, thanks to new conceptual systems allowing the contradictions to be outdistanced by integration of the data of these new systems.

Catharsis and memory

This process of awareness of the cognitive conscious recalls the one psychoanalysts describe as *catharsis,* which is both an awareness of the affective conflicts and a reorganization which enables the conflicts to be overcome. I am not competent in psychoanalysis and therefore prefer to remain careful, but it seems to me that *catharsis* is wholly different from a mere illumination, otherwise we would fail to understand its therapeutic action: It is therefore a reintegration and an elevation of conflicts thanks to a new organization. But where does this come from?

Erikson had a very interesting thesis in this respect: The affective present is well determined, as Freud shows, by the individual's past, but the past itself is constantly restructured by the present. This is profoundly true of the cognitive systems, and that is why awareness is always not only an expression or an evocation but also partly a reorganization.

But Erikson's argument consists of a choice between two possible interpretations of memory and, more specifically, the necessity of adopting the second. The first consists of considering memories as stocked (or accumulated) as such in the unconscious where evocation would remove them at will without modification or reorganization. The second interpretation, on the contrary, means admitting that every operation of the memory of evocation includes a reorganization; in other words, memory works in the manner of a historian who, studying some incomplete documents, in part deductively reconstructs the past.

The existence of false memories already is sufficient reason to justify the second interpretation, since they offer themselves to the conscious with the same characteristics of vividity or apparent reality as true memories. I myself, for example, have a very precise, very detailed, and very lively memory of having been the object of a kidnapping when I was still a child strapped to my carriage. I recall a series of precise details of the site of this adventure, the struggle between my nanny and the thief, the arrival of passersby, the policemen, etc. When I was fifteen, the nanny wrote to my parents that she had invented the whole

story and that she herself had been responsible for the
scratches on her forehead, etc. In other words, about the
age of five or six, I must have heard the story of this kidnapping which my parents then believed and, using this
story, I invented a visual memory which today still remains. This then is a matter of a reconstruction, although
false, and if the event had really occurred and consequently
if the memory were true, it is quite probable that I would
also have reconstructed it in the same manner, for no memory of evocation (but only one of recognition) exists for
a baby in its carriage.

Together with B. Inhelder and H. Sinclair, we have
tried to analyze the mechanism of memory during the
course of child development and we found the following
kind of facts which speak very clearly in favor of reconstruction. A child is shown a series of six rulers arranged
in decreasing size which he looks at for a moment without
himself arranging them. A week later he is asked what he
had seen and a certain number of memory levels are found:
(a.) a few small sticks all equal, (b.) large-small, large-
small pairs, etc., (c.) large-medium-small trios, etc., (d.) cor-
rect series but too short, and (e.) the whole series. Already
we see here that what is recorded in memory is not the
perceptive and objective fact except in (e.) but rather the
idea the child creates of it. After six months, (and naturally
without showing the model again) seventy-four percent of
the children showed slight progress in relation to the preceding memory: A few children of category (a.) went to
category (b.), a few (b.) into (c.), etc. Thus we can state

that the memory-picture merely forms a symbol representing a scheme (here a scheme of seriation): In six months, the scheme made progress and the symbol-image which expresses it must conform to its new form.

It is obvious that memory does not progress in this manner and that in the majority of cases disintegration occurs. But in all cases we observe some schematization which can serve to raise certain cognitive conflicts. In a diagram such as

the young child perceives the numerical equality of the upper and lower elements (four and four) and hence he concludes that the length of the two lines should be the same. But for him, the two lines of the same length should have the same extremities. Thus in his memory, the "W" is often completed by new strokes until the two far ends coincide. This is another example that shows the part of reconstruction in memory.

Thus all these facts are such as to make us very careful in utilizing childhood memories, for if, in the cognitive field, memory is already a more or less adequate reconstruction, it is obvious that, with the intervention of all kinds of affective operations, etc., the reconstruction will be all the more complex. Thus one should organize a whole comparative study on the affective and cognitive transformations of memory.

The problem of the stages

A series of studies has already been undertaken on the relations between our analyses of the cognitive development during the sensorimotor period and Freud's work, including the stages which he distinguished during the same period. I would like to recall, for example, Rapport's analyses and Wolff's fine work as well as Cobliner's appendix to Spitz's last work.[6]

In regard to the Freudian stages and my own, however, an important result was obtained by Gouin-Décarie in Montreal on the relation between the cognitive development of the scheme of the permanent object and the development of the *objectal relations* in the Freudian sense of the term. I previously showed that an object disappearing from a child's perceptive field is not conceived as permanent and is not sought for behind the screen where it has just disappeared. It is reabsorbed, so to speak, instead of continuing to occupy a position in space. Toward the close of the first year, however, as a function of these successive displacements, the child begins to seek it. Gouin-Décarie was able to show that this new reaction was a function of the construction of the objectal relations and that, in general, a correspondence exists among these forms of development. I had shown moreover that the first objects gifted with (cognitive) permanence were other persons and not inanimate objects, and Gouin-Décarie has

6 René A. Spitz, *De la naissance à la parole* (Paris: Presses Universitaires de France, 1968).

since verified this hypothesis (which had been based only on the observation of a single subject).

Other correlations can be observed between the cognitive and affective stages. At the age of seven or eight, for example, new reciprocity relations are developed (in the logical sense of the term) in connection with the formation of reversible operations. In the field of moral feelings, a lessening of the effects of the superego and of authority is noted on this same level in favor of feelings of justice and other aspects of the moral or affective reciprocity. Similarly, on the adolescent level, during the period when the individual inserts himself into adult life, relations exist among the affective and cognitive transformations.

From these multiple convergences, we would naturally not be able to conclude that it is the cognitive constructions which produce the affective modifications. But I do not believe any longer that it is the latter which determine the former, as psychoanalysts would be at first tempted to believe. (In this respect, see Odier's old research.) Affectivity or its privations can certainly be the cause of acceleration or delay in cognitive development, and Spitz showed this in his famous analyses. But this does not mean that affectivity produces or even modifies the cognitive structures whose necessity remains intrinsic. Actually the affective and cognitive mechanisms always remain indissociable although distinct; and it goes without saying that if the affective stems from an energetic then the cognitive stems from structures.

But the conclusion of all this is that many problems

remain to be solved and it is worth thinking, from now on, of the foundation of a general psychology based simultaneously on the mechanisms discovered by psychoanalysis and on the cognitive operations, for the kind of comparison we have made in the preceding pages is only a beginning which appears rich in promise.

3 The Stages of Intellectual Development in the Child and Adolescent

The stages of intellectual development form a privileged case and we cannot generalize to other fields. If for example we take the development of a child's perception or the development of language, we observe a completely different and much greater continuity than in the field of logico-mathematical operations. In the field of perception, in particular, I would be unable to give you a chart of stages similar to the one I offered you from the point of view of intellectual operations, for we rediscover this continuity from the organic point of view, a continuity which can be broken up in a conventional manner but which does not offer very distinct natural breaks.

To the contrary, in the field of intellectual operations

we witness this dual phenomenon: On the one hand, we see structures form which we can follow step-by-step from the earliest main features and, on the other, we witness their completion, that is, the formation of levels of equilibrium. Take for example the organization of whole numbers: We can follow this structuration based on numbers 1, 2, 3, and so forth, until the moment when the child discovers the series of numbers and at the same time the first arithmetic operations. At a given moment, this kind of structure is thus formed and leads to its level of equilibrium. This equilibrium is so stable that the whole numbers will no longer be modified during an entire lifetime while integrating themselves into more complex systems (fractional numbers, etc.). We are therefore in the presence of a privileged field in the heart of which we witness the formation of structures and their completion, where different structures can succeed one another or integrate themselves according to multiple combinations.

In this particular field and, I repeat, without my posing the problem of generalization, I shall call *stages* the breaks which fit the following characteristics.

1. If stages are to exist, first of all the *order of succession of acquisitions must be constant*. I speak not of the chronology but of the order of succession. The stages in a given population can be characterized by a chronology, but this chronology is extremely variable. It depends on the individual's previous experience, and not only on his maturation; and it depends above all on the social milieu which can hasten or delay the appearance of a stage, or

even prevent the manifestation. We find ourselves here in the presence of considerable complexity, and I would be unable to make a statement on the value of the average ages of our stages so far as any population is concerned. I consider only the ages relative to the populations which we have studied; they are therefore essentially relative. If it is a question of stages, the order of succession of conduct is to be considered constant; that is, one characteristic will not appear before another in a certain number of subjects and after another in another group of subjects. Here, where we witness such alternations, the characteristics at stake are not utilizable from the point of view of stages.

2. Then there is *the integrative characteristic,* that is, that the structures constructed at a given age become an integral part of the structures of the following age. For example, the notion of the permanent object which is constructed at a sensorimotor level will be an integral element of the notions of future conservation (when there will be conservation of a unit, or of a collection, or even of an object whose spatial appearance is distorted). Similarly, the operations which we will call concrete will constitute an integral part of the formal operations, in the sense that the latter will constitute a new structure but based on the former. (The second thus constitute operations carried out on other operations.)

3. Together with Mlle. Inhelder, we have always tried to characterize a stage not by the juxtaposition of foreign characteristics but by a *whole structure.* This notion acquires a precise meaning in the field of intelligence, more

precise than elsewhere. On the level of concrete operations, for example, a structure will be a grouping with the logical grouping characteristics found in the classification or in the successions. Later, on the level of the formal operation, the structure will be the group of the four transformations which I shall soon mention, or the network. Thus we mean structures which can be characterized by their laws of totality in such a manner that, once such a structure is achieved, one can determine every operation it covers. Considering that the child achieves this or that structure, it is thus known that he is capable of a multiplicity of distinct operations, and at first often without any visible relation among them. Here is the advantage of the notion of structures: When structures are complex, a series of operational schemes with no apparent connection among them can be reduced to greater unity; this is the whole structure as such which characterizes a stage.

4. A stage thus includes both a level of *preparation* on the one hand and of *completion* on the other. For formal operations, for example, the stage of preparation will be the whole period from eleven to thirteen or fourteen years, and the completion will be the level of equilibrium which appears at that time.

5. However, since the preparation of later acquisitions can involve more than a stage (with various overlapping among certain preparations, some shorter and others longer), and since there are various degrees of stability in the completions, in any series of stages there must be a distinction between the *process of formation,* or of birth,

and the *final forms of equilibrium* (in the relative sense). Only the latter constitute the whole structures mentioned in paragraph 3; whereas the formative operations are presented in aspects of successive differentiations of such structures (differentiation of the earlier structure and preparation of the following one).

Finally, I would like to insist on the notion of *lag* (of operations), to which we will return, for it can create an obstacle to the generalization of stages and introduce considerations of caution and limitation. Lags characterize the repetition or the reproduction of the same formative operation at various levels. We distinguish *horizontal lags* and *vertical lags*.

We will speak of horizontal lags when the same operation applies to different content. In the field of concrete operations, for example, a child aged seven or eight will know how to seriate quantities of material, lengths, etc. He will know how to classify, count, and measure them, etc., and will likewise acquire notions of conservation relative to these contents. But he will be incapable of all these operations in the field of weights, whereas on an average of two years later, he will be able to generalize them by applying them to this new content. From the formal point of view, the operations are the same in both cases but applied to different fields. In this case, we will speak of horizontal lag within the same period.

A vertical lag, on the contrary, is the reconstruction of a structure by means of other operations. At the close of the sensorimotor period, an infant achieves what with

H. Poincaré we call a *group of displacements:* The infant
will know how to orient himself in his apartment with detours and return, etc. But this *group* is only practical and
by no means representative. Several years later when it is
a matter of representing the same displacements, that is,
imagining or interiorizing them in operation, we rediscover similar stages of formation but this time on another
level, that of the representation. It is a matter of other
operations, and in this case we will call it vertical lag.

We will divide intellectual development into three
important periods.[1]

The period of sensorimotor intelligence

This first period extends from birth to the appearance of
language, that is, approximately the first two years of existence. We subdivide it into six stages:

1. *Reflex exercises:* From zero to one month.

2. *First habits:* Beginning of primary stable condition-
ings and circular reactions (that is, relative to the body
proper: for example sucking his thumb). From one to four-
and-a-half months.

3. *Coordination of vision and prehension* and beginning of secondary circular reactions (that is, relative to the
manipulated limbs). Beginning of coordination of qualitative spaces, until then undifferentiated, but without search-
ing for lost objects; and beginning of differentiation be-

---

[1] We will speak of *periods* to designate the important unities and speak
of *stages,* then of *substages,* to describe their subdivisions.

tween aims and means, but without referring to previous aims while acquiring new conduct. From about four-and-a-half to eight or nine months.

4. *Coordination of the secondary schemes* with, in certain cases, utilization of known means to achieve a new objective (several new means possible for one and the same goal and several goals possible for one and the same means). Beginning of search for the lost object but without coordination of successive displacements (and localizations). From about eight or nine to eleven or twelve months.

5. *Differentiation of schemes of action by tertiary circular reaction* (variation of conditions by exploration and directed groping) and *discovery of new means*. Examples: supporting devices (drawing a blanket to bring to the floor the object placed on it; negative reaction if the object is alongside or beyond the support), use of string or stick (by groping). Search for the disappeared object with localization in function of perceptible successive displacements and beginning of organization of the practical group of displacements (detours and returns in actions). From about eleven or twelve to eighteen months.

6. *Beginning of the interiorization of the schemes and solution of a few problems with action stopping and sudden understanding.* Example: direction of the stick when it has not been acquired by groping during stage 5. Generalization of the group practice with, incorporated into the system, a few nonperceptible displacements. From about eighteen to twenty-four months.

These six stages present a rather striking character if

we compare them to the stages of future representative thought, in the sense that they form a kind of prefiguration according to the term cherished by our president Michotte (analogous in a sense to the prefiguration of the notional, which he often mentions in regard to perception). Actually, on this practical level, we are witnessing an organization of movements and displacements which, first centered on the body itself, gradually decentralize and lead to a space in which the child situates himself like an element among others (hence a system of permanent objects including his body on the same ground as the others). On a small scale and on the practical level, we see here exactly the same operation of progressive decentration which we will then rediscover on the representative level in terms of mental operations and not simply of actions.

The period of preparation and of organization of concrete operations of categories, relations, and numbers

We will call concrete operations those we bear on manipulable objects (effective or immediately imaginable manipulations), in contrast to operations bearing on propositions or simple verbal statements (logic of propositions).

This period extending from about two to eleven or twelve years can be divided into a subperiod *A* of functional preparations of the operations [2] but of preoperatory

---

[2] If we call *operations* the interiorized, reversible, and solidary actions of whole structures such as the *groupings, groups*, and *networks*.

structure, and a subperiod *B* of operatory structuration itself.

THE SUBPERIOD OF PREOPERATORY REPRESENTATIONS

This subperiod can itself be subdivided into three stages:

1. From two to three-and-a-half or four years: *appearance of the symbolical function and beginning of the interiorization of the schemes of action in representations.* This stage is the one on which we have the least information on the operations of thought, for it is not possible to question the child before the age of four in a continuous conversation; but this negative fact alone is a characteristic indication. The positive facts are the following. (1) The appearance of the symbolical function in its various forms: language, symbolic play (or imagination) in contrast to the simple play of exercise, postponed imitation, and probably beginnings of the mental image conceived as internal imitation. (2) Plan of the nascent representation: difficulties of application to nonproximate space and to nonpresent time of the schemes of object, space, time, and causality already used in the effective action.

2. From four to five-and-a-half years: *representative organizations founded either on static configurations or on an assimilation to the action itself.* Duality of states and transformation is the characteristic of the first representative structures revealed on this level by questions about objects to be manipulated. The first are thoughts as configurations (compare the role of perceptive configurations,

figural collections, etc., to this level of nonconservation of totalities, quantities, etc.), and the second are thoughts assimilated to actions.

3. From five-and-a-half to seven or eight years: *articulated representative regulations.* Intermediary phase between nonconservation and conservation. Beginnings of connection between states and transformations, thanks to representative regulations enabling these to be thought in semireversible forms. (Example: increasing articulations of classifications, relations of order, etc.)

THE SUBPERIOD OF CONCRETE OPERATIONS

This stage extending from seven or eight to eleven or twelve years is characterized by a series of structures on the point of completion which can be carefully studied and their form analyzed. On the logical level, they all amount to what I called *groupments,* that is, they are not yet *groups* nor are they *networks* (with lack of boundaries for some and greater boundaries for others, these are half-networks): such are the classifications, the successions, the correspondences point-by-point, the simple or successive correspondences, the multiplicative operations (matrices), etc. On the arithmetic level, I will add the additive and multiplicative groups of whole and fractional numbers.

This period of concrete operations can be divided into two stages: one of simple operations and the other of completion of certain whole systems especially in the fields of space and of time. In the field of space, it is the period when the child of only about nine or ten reaches the systems of coordinates or of references (representation of verticals and horizontals in relation to these references). This is also the level of the coordination of whole systems of perspectives, the level which marks the greatest system on the concrete plane.

The period of formal operations

Finally there is the third and last period, that of *formal operations*. Here, as early as eleven or twelve (first stage) with an equilibrium level about thirteen or fourteen (second stage), we witness a great many transformations which are relatively rapid at the time of their appearance and extremely varied. We were able to reach these conclusions chiefly through the fine studies made by Mlle. Inhelder on inductive reasoning and on the experimental method among children and adolescents. At that age operations appear as different from one another as the following. First combinative operations: Until then there was only simple interlocking of sets and elementary operations, but not what mathematicians call *sets of parts,* which are the starting point for these combinations. On the contrary, the combination begins at about eleven or twelve and creates the network structure. On this same level, proportions appear, as well as the capacity for reasoning and self-representation according to two reference systems at the same time, the structures of mechanical equilibrium, etc. Let us study, for example, the relative movements of a snail on a small plank which will be moved in a direction opposite the snail, and the calculation of the result of these movements, one in relation to the other and in relation to a system of external reference. In this case we see (and they are found again in the mechanical equilibriums, etc.) the intervention of four coordinated operations: a direct operation (I) and its opposite (N), but also the direct operation and the opposite of the other system which formed the reciprocal of the first (R) and the negative of this reciprocal or correlative $(NR = C)$. This group of four transformations $INRC$ appears in a series of different fields in these logico-mathematical problems but also in problems of proportions even independently of school knowledge.

Above all, what appears in this last level is the logic of propositions, the capacity to study statements and propositions and no longer only objects placed on the table or immediately represented. The logic of statements also supposes the combinative network and the group of four transformations $(INRC)$, that is, the two complementary aspects of a new whole structure taking in every operatory mechanism which we see formed at this level.

I will end by saying that these three important periods with their particular stages form operations of successive equilibrium, steps toward equilibrium. The moment the equilibrium is reached on a point, the structure is integrated into a new system being formed, until there is a new equilibrium ever more stable and of an ever more extending field.

It is worth recalling that equilibrium is defined by

reversibility. To say that there is a step toward equilibrium means that the intellectual development is characterized by a growing reversibility. The reversibility is the most apparent characteristic of the act of intelligence which is capable of detours and returns. Thus this reversibility increases at a regular rate, level-by-level, during the course of the stages which I have just briefly described. It is presented in two forms: one which can be called inversion or negation which appears in schoolroom logic, arithmetic, etc.; the other, which we could call reciprocity, appears in the operations of relations. In every level of concrete operations, inversion and reciprocity are two processes traveling side-by-side and at the same time, but without meeting in a unique system. With the group of four *INRC* transformations, on the contrary, we have inversion, reciprocity, negation of the reciprocal, and the same transformation, that is, the synthesis in a single system of these two forms of reversibility until then parallel but with no connection between them.

In this privileged field of intellectual operations, we thus arrive at a simple and regular system of stages, but it is perhaps characteristic of such a field of perception that I am unable to furnish its stages.

4 *Child Praxis*

Praxis or action is not some sort of movement but rather a system of coordinated movements functioning for a result or an intention. To take but one example, the displacement of an arm which interferes in the act of putting on or of removing a hat is not praxis; a praxis consists of an action in its totality and not of a partial movement within this action. Praxis is an *acquired* as opposed to a reflex coordination; this acquisition can derive not only from the child's experience or education in the large sense (instruction, example, etc.), but also eventually from the internal operations of equilibrium which expresses a regulation or a stabilization acquired from coordination.

Thus characterized, praxis consists of two possible forms of coordination, the first constantly at work and the second capable of superimposing itself upon or of deriving from the first. We will call the first *internal coordination,* that which gathers several partial movements into a whole act, whether some of these partial movements previously existed in an isolated state (which is not the rule but can be observed), whether coordinated for the first time, or even whether the result of a progressive differentiation during gradual coordinations. We will call *external coordinations* the coordinations of two or several acts of praxis into a wholly new praxis of superior order, the earlier ones remaining capable of functioning in separate states.

The psychological problems which are now posed and which become directly or indirectly part of the clinical problems of apraxia, can be grouped into three major headings: (1) those of the mode of coordination (internal or external) peculiar to praxis; this will lead among others to choosing among the explicative models such as the associationist ones, the gestalts, or the assimilative schemes; (2) those of the relations between the coordinations proper of praxis and intelligence; and (3) those of relations between these coordinations and the symbolical function, particularly the mental picture. The problems in (2) are themselves subdivided into two groups. On the sensorimotor levels of development, that is, before language, the question arises if, specifically, the coordinations of praxis are gradually directed by an outer practical intelligence (in this case, the contents of this intelligence should be discovered) or if, on the contrary, the sensorimotor intelligence is nothing more than the very coordination of the actions. The problems in (2) are therefore the same as those in (1) on these initial levels of the development. After the construction of the symbolical function, the problems in (2) make one wonder what are the relations between praxis and the fundamental mechanism of representative intelligence, that is, the mechanism of the *operations*; these to be conceived precisely as actions of a certain kind, interiorized actions coordinated in well-defined structures (logico-mathematical structures, chiefly geometrical, etc.). As for the problems in (3), they are partially independent of those in (2), if we admit that knowledge or gnosis includes two distinct aspects: the operative aspect to which I just alluded, and the figurative aspect (perception, mental picture, etc.) intervening among others in the symbolical function in regard to the elaboration of the significants or symbolizers (for example, the picture).

To deal with the three kinds of problems, we are first going to study the sensorimotor levels, then the relations between praxis and the operations of representative intelligence, and finally, the relations between praxis and mental pictures.

Between the newborn child's almost entirely reflex behavior (but with the diffused cortical control emphasized by Minkovski) and the appearance of language or of the symbolical function, there exists a series of levels whose very succession is already instructive in the modes of coordination which characterize praxis and its relation to intelligence.

In the first of these stages, certain complex reflexes, like those of sucking, give rise to a kind of exercise and of internal consolidation due to their functioning, which announce the formation of schemes in behavior.

We call *schemes* of an action the general structure of this action, conserving itself during these repetitions, consolidating itself by exercise, and applying itself to situations which vary because of modifications of milieu. In this respect, sucking reflexes create a scheme (which is not the case of all the other reflexes but only of some of them) which manifests itself among others by the functional consolidation I just mentioned, but also by a certain number of generalizations (empty sucking, sucking any object in the presence of the lips) and of recognitions (finding the nipple again when moving slightly aside and distinguishing it from the surrounding teguments, etc.).

As early as the second stage, the presence of such schemes permits certain new acquisitions (new in relation to the original hereditary structures), thanks to the incorporation of new elements into the initial circuit: After sucking his thumb during fortuitous contact, the baby will be able, first, to hold it between his lips, then to direct it systematically to his mouth for sucking between feeding. Already we are in the presence of a praxis.

With the third stage, marked by the coordination of vision and prehension (coordination, according to Tournay, due to a myelinization of the pyramidal fasciculus, but which in addition requires an undeniable part of exercise [1]), the possibility of thus intentionally seizing the objects appearing in the close visual field creates the formation of a series of new schemes.[2] To mention only one, a child seizes among others a cord hanging from the top of its cradle; this shakes the top with all the objects which we had hung there (celluloid dolls filled with granules to produce sound, etc.). Soon afterward, when the objects had been removed from the top, we attached another object and the child, having watched this, at once sought the cord and pulled it while again watching the object hanging there. Subsequently, the balancing of a presented object from three to six feet from the cradle and even the interruption of repeated whistling gave rise to looking for and pulling the cord.

During a fourth stage, the child no longer limits himself to reproducing the sequences discovered by chance (circular reactions), but he uses the schemes thus discovered by coordinating them, one of these schemes assigning a goal to the action and one of the others serving as a means

---

[1] With our three children, this coordination was formed at six months, four-and-a-half months, and three months and three days, hence with considerable age difference but in similar order to the whole context of their activities.

[2] Beginning with the scheme of intentional prehension itself, quite distinct from the earlier reflex prehension due to the fact that the intentional prehension includes the possibility of "releasing," that is, choosing not to take. (This is not to be confused with the scheme, appearing much later, of purposely dropping an object from his hands.) The difference between this intentional prehension with the possibility of not taking is, as Ajuriaguerra pointed out to us, comparable to the active visual exploration (fixations and displacements of the intentional glance) as opposed to the roaming and gripping glance.

of achieving the goal. Or again, by presenting a new object, the child applies to it in turn (as exploration) each of the known schemes, in order to determine the practical significance or the use of this object, and he will grasp it to look at it, to suck it, etc.; he will shake it, rub it against the side of the cradle, hit it with one hand while holding it with the other, etc. In short, the stage is characterized both by a growing mobility of the schemes of action and by the appearance of what we called earlier the external coordination between acts of praxis.

During the fifth stage (beginning of the second year), the external coordinations are accompanied by a differentiation of the schemes as a function of experience; for example, reaching an object too far away by pulling the support (carpet, etc.) on which it had been placed, with variations in accordance with the situations. Here, therefore, there is simultaneous external coordination of schemes capable of functioning separately and there is discovery of new means by accommodating the schemes to the unexpected facts of experience.

Finally, during the sixth stage, which coincides with the first manifestations of the symbolic function, a beginning of interiorization of the external coordination between the schemes is manifested in the form of insight or of invention of new means. One of my children, for example, in order to reach an object placed in a slightly opened box of matches, began by feeling the box in various ways (fifth-stage behavior). Then after a pause during which he had observed quite carefully the much-too-small opening, he slipped his finger into this opening and thus solved the problem. This beginning of interiorization of the coordinations is often accompanied by symbolical gestures favoring the formation of the nascent representation. Thus while glancing at the opening which he wishes to enlarge, this child opened and closed his mouth several times, not because the coveted object in the box was to be eaten (he saw that this was a thimble), but more probably to symbolize the desired solution (to increase the opening).

Such being the stages of formation of elementary praxis peculiar to the sensorimotor period of development, let us now ask ourselves of what do the coordinations which characterize them consist.

First, it is worth noting that such development could not be reduced to an associationist model by learning or by conditioning interpreted in the sense of associations. Indeed, a scheme is more than a mere "hierarchical family of habits" (in the Hull sense) due to cumulative associations, for a new acquisition consists not only of associating a new stimulus or a new response-movement to stimuli or to previous movements *a, b,* and *c.* Any new acquisition consists of *assimilating* an object or a situation to a previous scheme by thus enlarging it. It is insufficient, for example, to explain the habit of thumb sucking by saying that the infant has *associated* his thumb to a sucking movement, for the real conditioning problem is to know why it stabilizes itself, when, like any association, it is merely of a temporary nature. Indeed, the thumb stimulus releases the sucking response only if it assumes a significance as a function of the scheme of this response, that is, if it is assimilated as a sucking object. Psychoanalysts would simply say that it is a breast symbol, but this apparent simplicity consists of attributing to the subject somewhat too precociously the very complex symbolical function.[3] Let us therefore satisfy ourselves by saying that it is assimilated to a sucking scheme and let us try to specify the meaning of these terms.

Assimilation thus understood is a very general function presenting itself in three nondissociable forms: (1) functional or reproductive assimilation, consisting of repeating an action and of consolidating it by this repetition; (2) recognitive assimilation, consisting of discriminating the assimilable objects in a given scheme; and (3) generalizing assimilation, consisting of extending the field of this scheme. Hence assimilation, on the behavior level, is merely the continuation of the biological assimilation in the large sense—any reaction of the organism to the milieu consisting of assimilating the milieu to the structures of the organism. Just as, when a rabbit eats cabbage, he is not changed into cabbage but, on the contrary, the cabbage is changed into rabbit, so in all action or praxis, the subject is not absorbed in the object, but the object is used and "included" as relative to the subject's actions.

It is therefore assimilation which is the source of

---

[3] One could, it is true, limit oneself to saying that the thumb = pleasure = breast. But this amounts exactly to what we call the assimilation of the thumb to the sucking theme, any assimilation being both cognitive (utilization or comprehension) and affective (satisfaction). In this respect, see the following paragraph.

schemes, with the exception of the orginal reflex and hereditary schemes which orient the first assimilations: Assimilation is the operation of integration of which the scheme is the result. Moreover it is worth stating that in any action the driving force or energy is naturally of an affective nature (need and satisfaction) whereas the structure is of a cognitive nature (the scheme as sensorimotor organization). To assimilate an object to a scheme is therefore simultaneously to tend to satisfy a need and to confer on the action a cognitive structure.

Thus, what we have called internal coordination of schemes is therefore nothing more than the product of cumulative assimilations. As for the external coordination between schemes, it is a question then of reciprocal assimilation. For any object, for example, capable of being seen (compare oculo-cephalogyric reflexes) and seized the coordination of the vision and the prehension includes a reciprocal assimilation of the corresponding schemes, the object becoming *both* something to see and to seize.

Thus conceived, the schemes of assimilation are not confused with the gestalts, although in certain cases, a scheme can present gestalt characteristics. A gestalt is an organization obeying laws of compensation or of intrinsic and independent equilibrium of the acquired experience: symmetry, regularity, simplicity, etc. Thus a scheme can obey the gestalt laws (symmetrical arm movements, etc.). The organization of a scheme, however, is much greater and results both in the subject's activities (which are a function of the utilization as well as of the laws of "good form") and of his acquired experience (accommodations to objects). The laws of compensation and of equilibrium of schemes stem therefore from the activities like those of the subject (to compensate an outer disturbance in order to satisfy a need, etc.) and not the so-called preformed geometrical laws.

It thus becomes relatively easy to solve the problem of the relation between sensorimotor praxis and intelligence (problems on which the interpretation of the ideomotor apraxia partially depend). If the mode of coordination of actions is truly of an assimilating nature and not simply associative, it becomes futile to subordinate the actions or praxis to a so-called intelligence which would be external to them and would consist then of a kind of faculty difficult to understand, unless of the first fact. There certainly exists a sensorimotor intelligence, and as early as the fourth stage, the mobility and external coordination of the schemes lead to a subordination of the means to the goals; the characterizing of this as intelligent action could not be denied (and this *a fortiori* with the discovery of the new means of the fifth stage and the insights of the sixth). But this intelligence is nothing more than the very coordination of the actions, and as early as the most elementary actions, we again find in the assimilation a kind of sketch or prefiguration of judgment: The infant who discovers that an object is to be sucked, to be balanced, or to be pulled orients himself in an uninterrupted line of assimilations leading directly to the superior behavior which the physicist uses when he assimilates (he also!) heat to movement or a scale to a system of work.

That is why as early as the sensorimotor praxis, the substructures of subsequent knowledge are outlined. The search for disappearing objects (long impossible, then developing gradually) leads to the scheme of the permanence of objects, which is a point of departure for the subsequent notions of conservation. The displacements in space are gradually organized into a scheme which takes form from what geometers call a group of displacements, and this scheme, already almost reversible [4] at the fifth and sixth stages, will play an important role in the organization of representative space, once reconstructed on the level of thought by interiorization of the actions in operation. The causality, the temporal series (order of succession, etc.) are not imposed on praxis from without by intelligence but develop under the effect of their coordination and constitute the substructures of the subsequent notions of cause, order, time, etc.

Let us now study child praxis as developed after the construction of the symbolical function, noting especially those whose disturbances correspond to what is known as constructive apraxia.

The symbolical function results from a differentiation between the significants and the signified (until now undifferentiated, as in the case of the perceptive signs or signals of conditioning). The symbols and signs, once differentiated from their significations, make it then possible to

---

[4] In the mathematical sense of the word and not neurological. Indeed, a group consists of the direct, reverse compositions (returns) and identical and associative ones (detours).

evoke objects and situations actually nonperceived, forming the beginning of representation. The significants,
which differentiate between one or one-and-a-half and two
years, so far as symbols proper are concerned, are: (1) symbolical play (representation of objects and actions by
gestures, etc.) dissociating itself from the mere play of
functional exercise; (2) postponed imitation (with its multiple varieties leading to graphic imitation or drawing);
(3) mental pictures doubtless resulting from interiorized
imitations. At the time when these various categories of
symbols are formed, there is also acquisition (by imitation,
etc.) of the systems of social signs, the principal one being
speech.

Thus the symbolical function makes this interiorization of actions possible or at least strengthens it considerably. We noted the beginnings at the sixth sensorimotor
stage: In addition to their material and effective development, the actions become more and more capable of being
carried out in thought or symbolically. This interiorization, however, supposes a reconstruction on the level of
thought, which is long and laborious: It is one thing for
the child, for example, to coordinate his displacements in
a group, enabling him to find himself again in his garden
or between his house and school, and another to be able to
represent these displacements in thoughts, respecting the
group rules (returns and detours, etc.), and to outline
these paths by drawing, language, or simply by arranging
the paths and the house on a model prepared for this pur-
pose. It is only after the age of seven or eight that represen-

tation rediscovers this group structure which was already active in the sensorimotor organization at the fifth and sixth stages.

Thus interpreted, representation or representative thought consists of two different aspects which should be clearly distinguished if we wish to state with some rigor the nature of the psychological trouble intervening in a constructive apraxia: the figurative aspect and the operative aspect.

The figurative aspect of thought is everything related to the configurations as such, in opposition to the transformations. Guided by perception and supported by the mental picture, the figurative aspect of representation plays an important role (abusively important and precisely at the expense of transformations) in the preoperatory thought of the child aged two to seven, before the operations are constructed in the sense I have just defined. Thus, when we pour a liquid from container A into a narrower and taller container B, the child aged four to six generally still believes that the quantity of the liquid increases because the level is higher. He reasons thus only on the configurations A and B by comparing them directly without the intermediary of the system of transformations (which would offer him the relation: higher but narrower; therefore equal quantity). After the age of seven or eight, on the contrary, he believes in the conservation of the quantity of liquid because he reasons on the transformation and subordinates the configuration to it.

The operative aspect of thought relates to transformations and is thus related to everything that modifies the object, from the moment of the action until the operations. We will call operations the interiorized (or interiorizable), reversible actions (in the sense of being capable of developing in both directions and, consequently, of including the possibility of a reverse action which cancels the result of the first), and coordinated in structures, known as operatory, which present laws of composition characterizing the structure in its totality as a system. Addition, for example, is an operation because it stems from collecting actions, because it includes a reversal (subtraction), and finally because the system of addition and subtraction includes laws of totality.[5] The operatory structures, for example, are the classifications, seriations, correspondences, matrices, series of numbers, spatial metrics, projective transformations, etc. A large number of logical, mathematical, and physical operations develop for the most part spontaneously in the child aged six or seven and are completed as of the eleventh or twelfth year by propositional or formal operations, making the adolescent's hypothetico-deductive deduction possible.

If we admit this distinction between the figurative and operative aspects of thought, it is then immediately evident that operations stem from the sensorimotor schematization, even if the symbolical function and the figurative representation are required for their interiorization and expression. Indeed, it should be well understood that an operation is not the representation of a transformation;

---

[5] Group laws, etc.

it is, in itself, an object transformation, but one that can be done symbolically, which is by no means the same thing. Thus an operation remains an action and is reduced neither to a figure nor to a symbol.

Thus the essential problem of praxis interpretation is to dissociate what is due to the figurative aspect and to the operations as such, in such a manner that in a constructive act of praxis, for example, we can diagnose what stems from intelligence or what stems only from the symbolical figuration.

Particularly in regard to space and the spatial disorders so important in apraxia, it should be understood above all that the spatial relations simply "given" in appearance between the external objects are in no way reduced in point of fact to pure systems of perceptions or of imagined representations but include operatory constructions far more complex than they appear. Although there exist vertical and horizontal positions locatable by postural and proprioceptive means and although directions can be estimated visually in relation to that of the glance (Donders' law), the prevision of the horizontal level of water in a slanted jar, for example, is not accessible to a normal child until about the age of nine, because it supposes a whole system of references bound up with Euclidian metric operations—the axes of coordinates capable of being constructed on the representative level only at the end of the long formation of measure operations. Even the conservation of lengths and distances, in a case of change in the arrangement of objects, is only acquired as a function of reversible operations, and is by no means acquired by the merely perceptive method or by the play of mental pictures only.

We must therefore turn to a precise investigation of the eventual operations in play if we wish to understand the details of troubles in a constructive act of praxis. Thus in the drawings of small bicycles so suggestive (and so similar to those of children aged five to six) furnished by Hécaen, Ajuriaguerra, and Massonnet in the case of right lesions,[6] we can ask ourselves to what extent the lacunae are due to the relations of causality, temporal series, spatial representations as figurative, topological relations (the chain "enclosing" the indented wheel, etc.), the absence of coordinates in the plan, etc. When we are told that "the copy of the Rey complex figure proves very defective," is this due to perception, to the graphic quality as such, or to many spatial operations which intervene implicitly in the success of this excellent overall proof but which cannot serve to dissociate the operative from the figurative aspect of the operation in play?

To specify the eventual relation between operations and praxis, it is now worth adding that, during the development of thought, operations go through three successive stages. During the first, between the ages of two and seven or eight, thought remains preoperatory, in the sense that the operations gradually form themselves but without achieving logical reversibility or adequate total structure, and they remain dominated by the figurative aspect of the representations.

---

[6] See Ajuriaguerra and Hécaen, *Le cortex cerebral*, 2nd ed. (1960), p. 270.

In the second stage (from seven or eight to eleven or twelve), certain operations are completed and organized in logically reversible structures. But (and this is important to the praxis problem) the operations remain concrete in the sense that they are limited to the field of manipulation of the objects and do not yet include simply verbal manipulation on the hypothetico-deductive level. In regard to seriation, for example, a child aged seven to eight succeeds in arranging, according to their increasing size, a series of small rulers (between ten and sixteen and a half centimeters) one above the other and in arranging them without hesitation according to a method (first, the smallest of all, then the smallest of those which remain, etc.). This is a fine example of praxis of an operatory character. Similarly, at the age of nine to ten, he will be able to arrange distinct weights (with objects of equal volume), which constitutes another operatory praxis. However these concrete operations alone will not enable him to solve the Burt test which is based on the same operations of seriation but on a hypothetico-deductive level: Edith is more blonde than Suzanne. Edith is darker than Lili. Which of the three is the darkest? Finally, at about the age of eleven or twelve, the propositional or hypothetico-deductive operations are constituted which can function beyond any object manipulation and no longer concern praxis.

Moreover, it is essential to note that, in addition to the acts of intelligence proceeding by concrete operations and tending to solve a problem of truth (with a true or false solution), there exists a considerable set of acts of intelligence tending to solve purely practical problems (which solutions are expressed in success or failure). Such in particular are the behaviors studied by A. Rey in his work on *L'intelligence pratique chez l'enfant* [7] and by Bussmann in his volume on *Le transfert dans l'intelligence pratique de l'enfant*. [8] For example, the withdrawal of an object from a container by using various stems, etc., as intermediaries (an early study of this kind having been furnished by two German psychologists, Lippmann and Bogen, *Naive Physik*). It is a question, in this case, of praxis in the strictest sense of the term, since the aim of these actions is principally of a utilitarian nature (to achieve a material result) and no longer cognitive, as in the acts of classification, seriation, or correspondence. But the interest of the research by Rey, Bussmann, and others was precisely to show the close analogy between the child's failures or successes and the operations of his thought itself at the levels considered. One of Rey's goals was to control this kind of prelogic which we pointed out in the child in the verbal area, if it could be found again in the area of practical intelligence. In the preface to Rey's first work, we insisted on the parallelism obtained. We could insist even more so today since we no longer limit ourselves to using verbal methods and have revealed the late character of concrete operations, that is, of the logic of object manipulation. In the context of practical intelligence, our pupil Bussmann revealed the transitions which exist between the

---

7 Alcan (Paris: Presses Universitaries de France).
8 Delachaux et Niestlé.

sensorimotor assimilation and the particularly logical generalization. From the viewpoint of the interpretation of the varieties of apraxia, this continuity between the practical intelligence and the particularly cognitive intelligence, if we can so state (hence the system of the logico-mathematical or logico-physical operations), seems instructive to us by emphasizing the relation of praxis and gnosis, in other words, of the basic unity of action and of intelligence in its operative aspect.

There now remains to be examined the figurative aspect of knowledge and of actions, especially the problems of image and of symbolical behavior.

The classic theories of apraxia would consider acts based on pictures. A. A. Grünbaum, on the contrary, interprets pictures as deriving from acts. From the psychological viewpoint, he is unquestionably right, and psychologists (Lotze, Dilthey, and others) have long shown that the picture does not consist of a mere continuation of perception but that it includes a propulsive element (compare Morel's, Schifferli's, and Rey's work). From the electroencephalographic viewpoint, Gastaut observes the same beta waves during the mental representation of the bending of the hand as during effective bending; Adrian has made similar observations. Using electrographics, Jacobsen, Allers, and Schminsky observed light peripheral activities (movement outlines) during the representation of arm movements simultaneous with the activities registered during the act itself thus represented. In short, the picture and the figurative aspect of thought as well as the operative aspect of thought and the operations themselves derive from the sensorimotor activities. How then should one conceive of this dual relation while maintaining the distinction of these two figurative and operative aspects of all knowledge?

We have just seen that the essential mechanism of the sensorimotor intelligence consists of a schematizing assimilation, and it is from this that the subsequent operations of representative thought proceed.

A scheme of assimilation, however, is constantly submitted to the pressure of the circumstances and can differentiate in accordance with the objects to which it is applied. We will call *accommodation* [9] this differentiation of response to the action of the objects on the schemes, synchronized with assimilation of the objects to the schemes. Equilibrium can then occur between assimilation and accommodation: Such is the result of an act of intelligence. But a primacy of accommodation can also occur and, in this case, the action is modeled on the object itself—for example, when the object becomes more interesting than the assimilating use the subject can make of it. Such more or less purely accommodating behavior indeed forms what is known as *imitation* and we can follow stage-by-stage the progress of this imitation on the sensorimotor levels in close correlation with the progress of intelligence (or equilibrium between assimilation and accommodation).

---

[9] In analogy to what biologists call *accommodates*, that is, the distinct phenotypical variations of the genotypical characteristics.

Thus our hypothesis is that the figurative aspects of thought derive from imitation and that it is imitation which assures the transition of sensorimotor to representative thought by preparing its necessary symbolism. On the one hand, at the sensorimotor levels, there is only imitation to constitute a kind of representation by gesture (naturally quite distinct from representation in thought which will eventually derive from it). On the other hand, the advent of the symbolical function, that is, as we have seen the differentiation of the significants and the signified, is due precisely to the progress of imitation which first becomes capable of functioning in its deferred form [10] (deferred imitation already constituting a true representation). This furnishes to symbolical games (beginning about the age of one and a half) their entire gestural symbolism which, as we are going to see now, forms the point of departure of the mental picture as interiorized imitation.

As early as 1935 [11] we insisted on this role of transition between the sensorimotor and the representative played by imitation. H. Wallon brilliantly returned to this idea in *De l'acte à la pensée*, by emphasizing the importance of the postural system and attitudes in the birth of representation. Thus we agree with Wallon on this point, but we do not believe that this relation is valid only for the figurative aspect of thought, whereas the operative aspect (which constitutes the chief characteristic of acts of intelligence

---

[10] That is, basically as early as its beginning in the absence of the model (in opposition to the early imitations in the presence of the model and continuing during its absence).

[11] *La Naissance de l'intelligence chez l'enfant,* pp. 334–355ff.

in opposition to their symbolical expression) continues the energy as such.

To return to the image, we therefore suggest conceiving it as an interiorized imitation,[12] and all research that we were able to do and are still doing on the development of the child's mental pictures reveals to what extent imagery (pictures) remains static and short before completion of operations, and above all to what extent imagery remains subordinated to operations instead of preparing and directing them. It is surprising, for example, to note the child's difficulties at the preoperatory levels in imagining the stages of a curve's transformation (in the form of wire thread) into a rectilinear shaft, or the rotation of a shaft around a pivoting center, or of the gradual growth in height of a stem placed on another, then shifted, or of a cube slipping on another, etc., before the spatial operations are constructed, with the conservation of the sizes during displacements.

This duality of image and operation seems important to us for the study of apraxia. One of the classic tests of

---

12 The first reason is of a genetic order: The behaviors of the first eighteen months appear to reveal the absence of pictures until what can be called the played picture (compare the infant who opens and closes his mouth before increasing the opening of a half-opened box) and the interiorized picture. The other reasons are the following: A sonorous picture (evoking the sound of a word, a melody, etc.) is accompanied by a production sketch, like the representation of a gesture. A visual picture continues not the perception as receiver but the sensorimotor activity of exploration which imitates the object's silhouette. (Compare Morel's and Schifferli's experiments on the ocular movements accompanying the picture and parallel to those which intervene in the perceptive activities during the very presentation of the object).

apraxia, which consists of imitating a transitive act without an object present, is based on imitative representation of the act and not on its execution in an operative situation. It is only starting at a certain level that the imagined representation of the act can play a role in the improvement of its execution, and when it is a question of somewhat complex acts for their anticipation to be necessary to the success; but it is easy to furnish a series of examples of acts correctly carried out by the child, where their representation is defective. The most remarkable example of imagined (pictured) representation capable of offering a precise anticipation of the acts and even of substituting for them, is the spatial intuition of the geometer who succeeds with surprising mobility in imagining every possible transformation of a figure, whereas a nongeometer, whom Plato [13] in his *Republic* proscribed, "sees" only a few. This geometrical intuition, though developed to a certain degree in any normal subject beginning at the level of concrete operations, remains, as we have just seen, oddly static and unfinished prior to this level. There is nothing easier for a five-year-old child, for example, than to pivot a rod at 90° until at a horizontal position (one end being fixed). This child's drawing, however, will disclose only the extreme positions; he is incapable of representing the intermediary slanted positions. Similarly, a child aged four to five, taking the same way from home to school and back, will find a systematic difficulty in reproducing it (even in outline)

---

[13] Plato is said to have inscribed, "Let no man ignorant of geometry enter here," above the entrance to his Academy.——TRANS.

on a model and will be satisfied with motor memories ("I go like that, then I turn, etc."), indicating by gestures rectilinear movements, sudden turnings, etc., but without evocations of reference points or of the way as such.

Generally, spatial pictures are thus dependent on actions and operations and not the contrary, and the mathematician's geometrical intuition is only an internal imitation of the operations which he is capable of doing according to an increasingly refined logic. Certain acts probably suppose at almost every level an imagined anticipation, for example, of drawing (oriented by Luquet's "internal model"). But these are figurative acts, so to speak, (the drawing is a graphic imitation as part of imitation in general) and the rule does not seem valid to us for operative acts ($=$ transformation and not reproduction of an object).

As for the corporal scheme, unfortunately we were unable to do research on this subject and therefore cannot make a statement on the picture's role in the actions exercised on the body itself. But whether this role proves necessary or not, we would have to ask ourselves to what point the construction of this scheme is precisely not bound up with imitation itself which we studied in regard to the first eighteen months [14] (following P. Guillaume's fine work on the learned and not innate character of this imitation). The child, for example, long knows his face only tactually and does not place it in relation to the faces perceived visually on others: Until the age of one year a yawning by

---

[14] Jean Piaget, *La formation du symbole chez l'enfant*, eds. Delachaux et Niestlé.

others is not at all contagious (if the experimenter yawns noiselessly!). The errors committed are far more instructive than the successes: To a model opening and closing its eyes, the child will reply among other actions by opening and closing his mouth, etc. If the corporal scheme were partly constructed as a result of imitation, the relation between the picture (or interiorized imitation) and the act would create a particular problem in this delimited field which would thus be midway between operative situations (as the intuition of space of objects) and figurative situations (as drawing).

Frankly, there is still nothing more ambiguous than the notion of the corporal scheme, despite the fine work by Head, Bartlett, Pick, Schilder, Conrad, and many others. In concluding their excellent work on *Méconnaissances et hallucinations corporelles* (Masson, 1952), Hécaen and Ajuriaguerra summarized well the present state of the act, but what is quite clear is the absence of a somewhat systematic genetic study, despite work by Wallon, Zazzo, Lezine, and others. Thus for the present, we can only conclude with Schilder: If somatognosis includes a set of perceptive facts, especially proprioceptive, it supposes above all a spatial setting integrating into a functional whole our perceptions, our postures, and our gestures. Thus it is extremely probable that in this setting are integrated not only the contributions of the body itself, but also the almost constantly indispensable reference which is visual, auditive, and partly tactilo-kinesthetic knowledge (as during imitation learning) of others' bodies and of what is common to every human body (and perhaps even animals). This is why, in the present fragmentary state of knowledge, we would be inclined to believe that somatognosis is established between the elementary sensorimotor schemes (which include the knowledge of hands, etc., but not that of the whole body) and the truly figurative symbolical behavior (pictures, etc.) and, requires as still-to-be representative or symbolical figurative instrument only imitation itself, whose role precisely is to assure the connection between the body proper and that of others.

Moreover, there remains to be specified how far we should extend the notion of corporal knowledge. But if we go so far as to include the notions of left and right and their application to the bodies of others as well as to one's own body (see Head's proof and our results on the difficulty, before about the age of seven, to designate the experimenter's left and right hand when seated facing the subject), it will be important to recall that even relations of this kind, while including an operatory and logical aspect, become part of the reciprocity setting whose point of departure is again furnished by imitation (in a one-way or mutual direction).

If a conclusion is expected of us, we might end this report by seeking points of contact between such a study of praxis and the analysis of apraxia. Ajuriaguerra and Hécaen suggest a new classification of apraxia based on the following trilogy:

1. *Sensorikinetic apraxia* is characterized by an altera-

tion of the sensorimotor synthesis with automatization of the gesture, but with no trouble in representing the act.

2. *Somato-spatial apractognosis* is characterized by spatial disorganization of the relations between the body and the external objects with no sensorimotor troubles. It is a question therefore of the somatognostic troubles causing gesture disadaptations, including disturbances of left-right relations, certain dressing apraxia, etc. In addition, there are often visual perceptivo-motory alterations but without this necessarily signifying primary perceptive troubles.

3. *Apraxia of symbolical formation* is characterized by disorganization of symbolical and categorical activity (ranging from the agnosia of utilization to frequent trouble with verbal formation).

To compare this list with what we have seen of normal praxis, we at once note certain correspondence but also note that a question, rather a central one, remains.

These three categories of apraxia correspond closely to three genetic levels: sensorikinetic apraxia to the sensorimotor level; somato-spatial apraxia to an intermediary level between the elementary sensorimotor behaviors and the behaviors made possible by the symbolical function, the intermedial level whose point of departure we suggested is found in the behavior of imitation; apraxia of symbolical formation, finally, to the level characterized by representations in their dual figurative and operative aspect.

But the remaining question precisely relates to this dual aspect of representative thought: Does apraxia of sym-

bolical formation result from alterations of the operations as such or only from the gestural symbolism, imagined or even verbal, serving to represent them? We dislike
the term *categorical* used by Gelb and Goldstein, Wallon,
and others to designate the notional or conceptual settings
which correspond to the verbal settings, for in this language "the symbolical and categorical activity," as one says
too easily, seems to constitute only a one and the same
"activity" of which, in point of fact, the only "active" characteristic would be to allow abstraction! Certainly it is
conceivable that this is so, but our effort consists in doubt-
ing such unity. Believe me, thinking cannot be reduced to
speaking, to classifying into categories, nor even to abstract-
ing. To think is to act on the object and to transform it.
When an automobile breaks down, an understanding of
the situation does not consist in describing the engine's
observable failure but in knowing how to take it apart and
reassemble it. In the presence of a physical phenomenon,
comprehension begins only by transforming the facts in
order to dissociate the factors and to make them vary separately—an action not of categorizing but of acting to
produce and to reproduce.[15] Even in pure geometry, knowledge does not consist of describing the figures but of transforming them to the point of being able to reduce them
to basic groups of transformations. In short, "In the beginning was the Act," as Goethe said,[16] and the operation fol-

---

[15] To achieve the "mode of production of the phenomena," despite the
ban pronounced by Auguste Comte.
[16] *Faust*, Part One, Scene III——TRANS.

lowed! Thus it seems to us, there remains to be established with some care to what extent constructive apraxia, ideatory apraxia, and in general apraxia of symbolical formation concern only symbolizing, that is, representation of the gesture, the design, the picture, or even the language, or if they are related to the symbolized itself, that is, to actions and operations.

5 Perception, Learning, and Empiricism

The goal of the Center of Genetic Epistemology of Geneva is to try to examine by means of psychogenetic methods (or by methods more theoretical than experimental but directly rounding out psychological research [1]), a certain number of epistemological hypotheses which can be verified by cases. In this respect, it is important to attempt to prove the validity of the interpretations of empiricism on the two privileged grounds which are classically invoked for its justification: that of perception, which is supposed to furnish us with immediate knowledge of external reality, and that of learning, an operation which is supposed to lead to an acquisition of knowledge through experience

---

[1] For example, the information or chart theory used respectively in booklets III and IV of our *Etudes d'épistémologie génétique* (Paris: Presses Universitaires de France).

only. I would like to summarize here our results relating to these two points.

The general problem that we have asked ourselves during the past two study years can be stated as follows. From the empirical viewpoint, and especially from its revived and contemporary form which is that of logical empiricism, there are two forms of knowledge: (a) empirical knowledge furnished by experience (perception and learning) independent of any logic and prior to the logico-mathematical coordinations; (b) logico-mathematical knowledge consisting of coordinations after the event and connected in particular with the use of language. On the contrary, the hypotheses which we propose to prove state (a) that on every level (including perception and learning), the acquisition of knowledge supposes the beginning of the subject's activities in forms which, at various degrees, prepare the logical structures; and (b) therefore that the logical structures already are due to the coordination of the actions themselves and hence are outlined the moment the functioning of the elementary instruments are used to form knowledge.

The first method of attacking such a problem in the field of perception consists of discovering if pure findings exist in the form of a simple recording of the perceptive facts or if, as early as elementary perception, the finding presents itself in the form of recordings and inferences.

The problem is not new since, as early as the beginning of experimental psychology, Helmholtz caused a set of quasi-inferences to intervene in perception, but that was contradicted by Hering in the name of a so-called more exacting physiologism. We are witnessing today a return to Helmholtz, especially in American psychology with the new look of Krech, Portman, Bruner,[2] and others, and with the transaction theory of Ittelson, Kantril, and others. Special note should be made of the manner in which W. P. Tanner and his colleagues at the University of Michigan were able to renew the theory of perceptive thresholds by turning to the statistical theory of decision. According to these authors, the sensory fact is never recorded in the pure state but in liaison with "noises" relating to its physical and physiological context. It is recorded in such a manner that, in order to perceive its existence on the threshold level, it would be a question of dissociating the stimulant from the "noise," hence to "decide" for oneself, with the risks this includes in terms of gains or losses of information, and thus to devote oneself to an inductive preinference.

It was on genetic grounds, however, that we tried to reveal the role of the perceptive preinferences. A young child, for example,[3] is shown briefly two parallel rows of four coins, one being spaced out more than the other: The subject will then have the impression that the longer row has the more coins. The child is then shown the same two rows but in such a manner that the elements of one are

---

[2] See J. Bruner, *Les processus de préparation à la perception* in *Logique et perception*, booklet VI of *Les études d'épistémologie génétique* (Paris: Presses Universitaires de France, 1958).

[3] For this experiment as well as for similar ones, see J. Piaget and A. Morf, chapter III of *Logique et perception*, booklet VI of the *Etudes*.

connected to those of the other by lines, thus presenting a material connection of a character either (I) on a one-to-one basis or (II) not. (In the latter case, the first element of the first row is connected by two lines to two distinct elements of the second row, the second and third element of the first row are connected by a single line to the third and fourth elements of the second, and the fourth element of the first remains unconnected.) The youngest children, having no scheme of one-to-one correspondence, perceived an inequality of coins on the figure provided with lines I, as they had on the figure without lines. At a second level of development, the child perceives, on the contrary, the equality in I (but does not see it without the lines); he sees it also in II, being then satisfied by an overall correspondence and no longer one-to-one. At a third level, he perceives the equality in I but not in II. At a fourth level, he perceives it again in II by dissociating the perception of the coins from that of the lines. Such an experiment shows therefore that the same material facts (figures I and II) are perceived differently according to the schemes at the subject's disposal. The application of these schemes to the given fact then supposes the intervention of nonpresent elements in perception and, consequently, of inferences (let us rather say of unconscious preinferences) based on these elements, preinferences required to confer this or that significance on the given facts.

In such cases, therefore, we could not dissociate the findings of inference, the problem of their relations thus reposing within the very core of perceptions and not, as is generally imagined, only at the frontiers between perception (conceived as the prototype of findings) and notional interpretation (conceived as the only seat of inferential operations).

Is conceptual representation itself capable of modifying a perceptive process in the same preinferential sense? Bresson conceived in our Center an ingenious experiment consisting of showing children a number with its upper section concealed which corresponded to the numbers 1 and 7 in such a way that only the slanting of the vertical line makes it possible to decide if it is a 1 or a 7. The child perceived the figure at the end of a sequence of pairs (such as 65, 66, and 67) leading him to anticipate either a 1 or a 7, and the perception was expressed not only by a verbal reading but also by a reaction of adjustment making it possible to reproduce on an appropriate device the slant of the principal bar. Here again, perception is revealed modifiable as a function of the child's inferences or preinferences.[4] Therefore, Bresson constructed a fine probabilist scheme of the form of perceptive learning which consists of discriminating, in a constantly refined manner, the stimuli of neighboring sets, because of an increasing number of indicators. Founded on Hamming's theory of information and codes, Bresson's scheme offers us a very suggestive model of the liaisons between perception and logic (categories, relations, and inferences) at the core of this essential type of perceptive adaptation.[5]

---

[4] F. Bresson, chapter V of *Logique et perception.*
[5] Ibid., chapter IV: *Perception et indices perceptifs.*

In a more general manner, the author of this paper, with A. Morf's collaboration, sought to liberate the partial isomorphisms between perceptive structures and structures of categories, relations, and inference,[6] to draw the unnatural conclusion that logic would be preformed in perception but that perception would not function without the intervention of a sensorimotor scheme bound up with the whole action and which itself would then be a point of departure of subsequent logical structures. In effect perception as such would not account for the formation of any logical-mathematical idea (nor even for any physical idea, because every idea implies, in order to elaborate itself, the intervention of a logical mathematical body). On the contrary, all perception, without doubt even at the most elementary level of field effects, is structured by sensorimotor activities larger than itself and for which coordinations prepare logical structures.

Similarly, in perfecting an idea which I had long developed, I attempted to show [7] that in every field (perception and association) in which the subject acquires some knowledge by reading experience, this reading does not consist of cumulative recordings but in assimilations, that is, in incorporating the facts in organizing schemes, due not only to the subject's activities but equally to the object's characteristics. Thus, during short presentation periods (with tachistoscope: research made in collaboration

6 J. Piaget and A. Morf, chapter II of *Logique et perception*.

7 Jean Piaget, *Assimilation et connaissance*, chapter III of *La lecture de l'expérience*, booklet V of *Les Etudes d'épistémologie génétique*.

with V. Bang and B. Matalon), the optico-geometrical il-
lusions pass in general (and for certain points of centration) for a *maximum* of 0.1 to 0.5 seconds, which implies
the existence of at least two factors: one of recording
("meetings" between parts of the figure and those of the
receptive organs), and the other of relating ("couplings"
between "meetings"), one including a source of distortion,
the other a source of possible correlations. Even in these
situations where elementary contacts are observed between
subject and object, a somewhat complex model of assimilation should be substituted for a mere "reading."

This assimilation is particularly striking in the case of
the geometry of perception. We know that Luneburg, an
American mathematician and psychologist, believed he
could establish that binocular space with convergence and
vergence (and free eye movement) presents a Lobachevskian structure, recognizable among other ways by the perception of parallelism (between avenues of luminous
points in darkness) when it is in conflict with that of the
equidistance. A. Jonckheere, who had checked Luneburg's
experimental facts in London (with verification of the
measure of curves taken empirically), took up the question
again in our Center [8] by means of an original device: a
cube with wire thread ridges spinning in front of a mirror
which reflected its image in the form of a second cube spinning within the first. He then posed a new problem: that
of the relations between the sensory facts and positive

---

[8] A. Jonckheere, *Géometrie et perception*, chapter VI of booklet V of the *Etudes*.

judgment, particularly as regards the distortions or the apparent rigidity of the studied cube. Unfortunately, we know nothing of these sensory facts. All that we can say about this is that, if they consisted of a projection of the objective facts on the subject's visual "picture," the results observed would speak in favor, in this particular case, of Euclidian perception. (But this would not be so if the projection were done, for example, on a sphere corresponding to the visual field of movement.) But there is certainly complex assimilation or, if we prefer, expression of the fact in a perceptive structure, and if we compare the Lobachevskian structure of the perceptions obtained in Luneburg's arrangements to the Euclidian structures of the current representation, we see that the subject has at his disposal at least two geometrical structures. Such a result is then ruinous not only for apriorism (for if this space corresponds to an *a priori* form of the sensibility, a sole necessary form would have to be imposed) but also for empiricism (for lack of identity between the perceptive space and the space of the objects such that the latter is structured by the experimenter or by the physicist on small observation scales).

The few results obtained on the grounds of perception would have been incomplete without a parallel study on those of learning. Indeed, there are two ways of acquiring knowledge as a function of experience: either by immediate contact (perception) or by successive liaisons as a function of time and of objective repetitions (learning).

Naturally, what we noted in the perceptive field leads us to believe that the results would be the same so far as learning is concerned, but it would have to be verified (which we did during our third year of activity). The two problems which we then asked ourselves in this respect are the following: (a) Does a learning of logical structures exist and, if so, is it identical to that of any behavior or of any physical successions? (b) Does the learning of any structures itself include a logic or a prelogic inherent in the mechanisms necessary to its functioning?

As regards the first of these two problems, Morf turned to one of our old results concerning the quantification of inclusion in children aged five to six. To the question: Is there more B than A, if all the As are Bs and if all the Bs are not As?, the child of the preoperatory level generally fails to answer correctly,[9] for he is unable to compare the whole B to its proper A section: No sooner is the whole dissociated in thought than the section A is then compared merely to its complementary A' (or $A' = B - A$). Morf then sought to submit the children to various forms of learning, one consisting of counting the As and Bs on various successive examples (or to note the extensions of the As and of the Bs), the other of allowing the child freely to manipulate the collections, and the third of explaining the possibility of intersections (x can be both an A and a B). The results obtained are instructive: (a) The mere reading of the facts (quantity of As and of Bs) is not sufficient to

---

[9] Naturally, the question is posed on concrete examples such as the pictures of flowers: Are there more flowers (there) or more primroses?, etc.

bring about the learning of the inclusion $A < B$ and, in the best cases, only leads to the finding $A < A + A'$, but without the combination $A + A'$, in the child's eyes, equaling the whole category B, and this probably for lack of sufficient understanding of the *all* and of the *some*. (b) Free manipulation, on the other hand, leads in a certain number of cases to understanding of the inclusion $A < B$, learning then consisting of an operatory exercise itself. (c) Intersection can also lead to inclusion, the operatory structure $A < B$ being, in this case, learned on the basis of another operatory structure.

In all, this first research thus seems to show that the learning of the logical structure in play is carried out on the basis of other operations or sketches of operations and not on the basis of findings analogous to those in which the learning procedes from a physical law.

P. Greco's research on the inversion of inversion led to similar results. Three elements are set in the ABC order on a rigid stem slipped into a tube which, by means of an 180° rotation, creates the order CBA, and by two rotations returns to the original order, etc.[10] What children aged five to six found, that is, successive results obtained during each manipulation, led to a degree of learning of the inversions of inversions. But this learning is limited and fails to lead to the construction of the particularly operative structure: Indeed, it is a question only of a somewhat advanced articulation of the preoperatory intuition which the

---

[10] See Jean Piaget, *Les notions de mouvement et de vitesse chez l'enfant* (Paris: Presses Universitaires de France, 1946).

child already possessed. Consequently, here again, learning of the structure consists in exercising the existing outlines, for in order to use the results of the experience, they must be understood and in order to understand them when it is a question of a set organized according to a logico-mathematical structure (which here is a group structure of order 2), previously understood instruments must be used.

J. Smedslund studied the learning of conservation and of transitivity of weight. Having children aged five to seven note on a scale the conservation of weight during a modification of the form of a small ball of clay (Piaget's and Inhelder's previous research had revealed the generally late character of this invariant which, in seventy-five percent of cases, is only acquired about the age of nine), he thus obtained a very appreciable learning of conservation. He was unable, on the other hand, to induce any immediate learning of transitivity. But two other groups of facts classified these results. On the one hand, the children, after acquiring the conservation of weight, revealed some weeks later the acquisition of transitivity. On the other hand, during previous child examination (designed to separate those who already possessed conservation), a high correlation was found between the degree of proof of conservation and that of transitivity. Thus we can interpret these facts as follows: (a) the physical aspect of conservation of weight gives rise to easy learning, which *ipso facto* does not involve learning of conservation as a necessary and transitive structure; (b) the logical structures in play, especially transitivity, give rise only to a limited learning consisting above all of spontaneous and internal organization of the empirical fact.

Smedslund found no differences, in his experiments on the learning of the conservation of weight, between the mere changes in form of the balls of clay and the reactions to the situations where the child began to take part in additions or subtractions of parts before judging conservation in modifications without addition or subtraction. J. Wohlwill thought that this weak role of addition and of subtraction perhaps stemmed from the continuous character of the quantities in play, and he proposed to analyze the effect of an exercise of these additive operations on the conservation of a set of discontinuous elements. (He turned his attention moreover to problems of perception and of conception of number.) The experiment revealed the role of the exercise of additive operations on the learning of conservation of sets and of number.

In conclusion, this small amount of research on the learning of the logical structures reveals that learning certainly exists but in a form both limited and specific: limited because one obtains from the subjects only a certain progress in the construction of the structure in play (and a progress conforming to the order of stages observed in development in nonexperimental situations), but not by this whole structure (except when the subject was able to develop it by spontaneous exercise); specific because to "learn" a logical structure, the subject must utilize as previous conditions nonlearned sketches of this structure or of other structures which imply it. The learning of logical structures is based therefore on a kind of circle or spiral, which amounts to saying that structures constitute the product not only of this learning but also of an internal operation of equilibration.

The second of our problems is then necessarily posed, namely, to know if the learning of any structures also includes a logic or a prelogic indispensable to its functioning.

In this respect, B. Matalon studied the learning of aleatory successions as well as double rotations (*AA, BB, AA,* etc.). Although he has yet to finish this research, he was able to note the existence of an interesting development in children as they aged: Whereas the youngest above all concentrated on the succession of their own actions, the oldest showed a lack of focus on the direction of successive goals. Thus the learning is a function not only of these given successions and of their repetitions but also of the subject's action coordination, this coordination including, by its very nature, a certain logic (let us recall that our previous analyses [11] led us to situate the sources of the logic not, or not exclusively, in speech but in coordinations of actions).

Similarly, M. Goustard studied learning at different ages in a labyrinth situation analogous to those which he had used in animal psychology. He thus obtained, at various age levels, very different learning curves: No learning at the age of five (for certain situations), increasingly rapid learning from the age of six to twelve or thirteen (but with backwardness at the age of eight to nine), and immediate

---

[11] See booklet IV of *Les Etudes (Liaisons analytiques et synthétiques).*

understanding (insight) beyond. Learning is therefore a function of the logical instruments at the child's disposal: insufficient at the age of five, modifying itself about the age of eight (the backwardness observed at this level corresponds to a change of methods due to the appearance of the operatory symmetries) and, as early as thirteen or fourteen, giving way to an immediate deduction which replaces learning.

In short, the learning of any structures seems itself to include a logic inherent to its functioning, comparable at the beginning to this prelogic already at play in perception, then tending to join the inductive and deductive structures which finally supply learning as such.

From the epistemological viewpoint, all this research seems to us to lead to the following conclusions. First, it seems to interpret the logical structures as *a priori* forms, since learning and experience are required for their elaboration. It is a question, it is true, of a very special type of experience which, like the physical experience, does not include an abstraction based on the characteristics of the object but an abstraction based on the actions effecting these objects and on coordinations which connect these actions (logico-mathematical experience). The learning of logical structures is itself therefore of a special type, since it consists simply of exerting or of differentiating previously acquired logical or prelogical structures.

But, in the second place, such results also fail to conform to the empiricist interpretation for a number of reasons. The chief reason is that neither the analysis of perception nor that of learning in general ever places us in the presence of a pure recording of external facts, either in the form of a pure perspective finding (perception always includes a part of inference or of preinference), or in the form of a purely associative recording (learning always includes an assimilating operation which itself causes a logic or a prelogic to intervene). The basic relation of stimulus and of response, even if we retain such a term, as well as the associations of stimuli and of responses, could not be interpreted therefore in the sense of an exclusive submission of the subject to the object. Certainly this submission exists, and even reinforces itself during development, but it is possible only because of the intervention of coordinating activities peculiar to the subject and which, in the last analysis, constitute the deepest source of the logical structures. In short, the object is known only so far as the subject achieves action on it, and this action is incompatible with the passive character which empiricism, at various degrees, attributes to knowledge.

6 Language and Intellectual Operations

Some forty years ago, during my first studies, at a time when I believed in the close relation between language and thought, I scarcely studied anything but verbal thought. Since then, there has been the study of the sensorimotor intelligence before language, the results which Rey obtained in his analysis of *L'intelligence pratique chez l'enfant,* then the inventory of the concrete operations of categories, relations, or numbers (with their infralogical parallel in the field of spatial operations and of measure) which develop between the ages of seven and twelve, much earlier than the level of propositional operations (the latter alone being capable of bearing on simply verbal statements). All this has taught me that there exists a logic of coordinations of actions far deeper than the logic related to language and much prior to that of propositions in the strict sense.

Language probably remains less a necessary condition for the completion of logical structures, in any case at the level of these propositional structures, nor does it become a condition sufficient for their formation, and this even less so as regards the elementary logico-mathematical structures. It is on these language inadequacies that I will chiefly dwell, for even if all see the decisive importance of its contribution which I hope to recognize, all forget too often the part of actions and of operatory intelligence itself.

The principal operatory structures are, it is true, included in current language in a form which is either syntactical or inherent to the meanings (semantic). First, in regard to concrete operations which bear directly on objects (categories, relations, and numbers), the linguistic distinction of nouns and adjectives corresponds in general to the logical distinction of categories and of predicates. And, as a function of the meaning attributed to different nouns, any language consists of relatively elaborated classifications: Keeping to the current meaning of the words sparrow, bird, animal, and living being, the subject speaking can conclude that all sparrows are birds, that all birds are animals, and that all animals are living beings without the reciprocal being true, which constitutes a hierarchical interlocking of categories, that is, a classification. To state, on the other hand, that whales are both mammals and aquatic animals consists in expressing an intersection or multiplication of categories, a principle of the multiplicative classifications, and no longer simply additive. The terms grandfather, father, son, brother, uncle, nephew, etc., are sufficient to determine the structure of a genealogical tree or of counivocal multiplications of categories or of relations. The comparatives "greater than," etc., lead to seriations, etc., and the series of whole numbers are part of the current vocabulary. As for propositional or formal operations, language formulates the principal ones: the implication ("if . . . therefore"), the exclusive or nonexclusive disjunction ("either . . . or . . . "). And the possibility of reasoning on simple hypotheses, the attribute of these hypothetico-deductive operations, is precisely assured by such use of language. The syllogistic is expressed directly by adequate verbal forms, to the point where we can reproach Aristotle's logic for having been somewhat dominated by grammar. As for structures much too differentiated and refined to be expressed by current language, mathematicians and logicians created for their own use artificial or technical languages but which, psychologically, are still languages.

It is natural therefore that, not only for psychologists but also for epistemologists, theories be imposed which seek to reduce to a single language, from a simultaneously genetic and causal viewpoint, the set of intellectual operations, if not to speak of the whole thought (with the sole exception of mental pictures of kinetic or visual order). There is no need in a gathering of psychologists to recall in this respect the work and tendencies of the behaviorist ma-

terial stemming from Watson. But it would be interesting
to point out the whole convergence of these positions in
those of an epistemological school which first worked completely independently (at the time of the Vienna Circle),
and then maintained the closest relations with behaviorism
in the strict sense after the "Viennese" had emigrated to
the United States. R. Carnap, one of the founders of this
logical empiricism (or positivism), began by maintaining
that all logic consisted only of a general syntax in the linguistic sense of the term. Later, and parallel with Tarski,
he was led to add a general semantic, but this also by no
means enables us to cross the frontier of language. Finally
Morris revealed the necessity (not recognized moreover
by the whole school) of considering the operative characteristic of logic, to complete the logistic syntax and the semantic by a pragmatic one; but it is still a question of the
rules of utilization of language and in no way of a logic of
action. If we glance through the pages of the *Encyclopedia
of Unified Sciences* which forms the summary of logical
positivism, we cannot help but be struck by the insistence
of the school's logicians, linguists, and psychologists (yet
noting how far more subtle E. Brunswick remains as compared to his nonexperimentalist partners) who repeat in
emulation of each other that the mentalist concepts of
thought, etc., no longer correspond to anything whatever,
that all is language, and that access to logical truth is assured by no more than a healthy exercise of language.

*   *   *

These are psychological questions and, consequently, only experience can decide them. In this respect we must take notice of the following two groups of problems.

I. Language can constitute a necessary condition for the completion of logico-mathematical operations without being a sufficient condition for their formation. On this point, the genetic facts are decisive in enabling us to establish: (a) whether the roots of these operations are prior to language or are to be sought in verbal behavior, (b) whether the formation of thought is connected with the acquisition of language as such or of the symbolical function in general; and (c) whether in the child's mind verbal transmission is sufficient to constitute operatory structures or whether this transmission is only sufficient when assimilated due to structures of a more profound nature (coordinations of actions) not themselves transmitted by language.

II. As for considering language as a necessary condition (but not sufficient) for the constitution of operations, there remains to determine: (a) whether operations function only in their linguistic form or whether they stem from *whole structures* or dynamic systems nonformulated as systems in the current language itself (in opposition to the technical language); (b) whether, nevertheless, in the completion of these eventual operatory structures, the role of language then remains necessary in a constitutive sense or merely as an instrument of formulation or of reflection; (c) in the case where language plays a constitutive role, whether it is chiefly a system of communications with all that this would include of control rules and error precorrection, or whether the structures are pre-established in a ready-made language.

1. In regard to the problems in category I, we can already recall the following facts, subject to the set of unsolved questions and experiments which remain to be done and which I will stress at the close of this report.

(a) On the sensorimotor level prior to the appearance of language, we already see the development of a whole system of schemes which prefigure certain aspects of the structures of categories and relations. Indeed, a scheme is what is generalizable in a given action: For example, after having attained a distant object by pulling the blanket on which it had been placed, the child will generalize this discovery into using many other aids to draw closer many other objects in various situations. Thus the scheme becomes a kind of practical concept and, in the presence of an object new to him, the child will seek to assimilate it by applying to it successively every scheme at his disposal, as though it were a question of these definitions by use, characterized by the words "it is for . . ." on which Binet insisted at a much later stage.

In generalizing themselves, the schemes first constitute kinds of classifications: The same goal, for example, can correspond to several means capable of achieving it and equivalents among them from such a viewpoint, or else the same means can lead to several goals. The categories include a *comprehension* from the subject's viewpoint, that is, a set of common qualities on which the generalization is based; they include, on the one hand, an *extension* (the set of situations to which they apply), but from the sole viewpoint of the behavior observed by the experimenter and without the subject's being capable of representing it, as he will succeed when he will have achieved the level of the symbolical function.

The schemes naturally include, on the other hand, a great variety of carrying out of relations, preludes of the logic of the relations which will subsequently develop on the representation level. These relations can even lead to kinds of sensorimotor seriations, as in the increase of plotted points of decreasing size (see Bühler's baby tests).

The coordination of schemes leads, moreover, to practical inferences: Seeking an object beneath a cloth under which a beret had been placed and not seeing the object when he raises the cloth, the infant sixteen to eighteen months old at once concludes that the object is beneath the beret, since this object had been slipped beneath the cloth and that in raising the cloth he fails to see it.

But the sensorimotor schematization leads above all to prefigurations of future notions of conservation and of the future operatory reversibility. Thus, between the middle of the first year and that of the second, there is developed this elementary form of conservation which is the scheme of the permanent object. This scheme already consists of a kind of group invariant: In effect, the search for the disappeared object is a function of its localization and the localizations themselves are only assured by the construction of a group of the displacements coordinating the detours (associativity of the group) and the returns.

We are thus led to admit that, prior to the operations

formulated by language, there exists a kind of logic of co-ordinations of actions, including, notably, relations of order and of interlocking liaisons (relations of the part to the whole). If, on the other hand, we distinguish at the core of the representations and of subsequent thought a figurative aspect linked to the representation of the states, we cannot help establishing a relation of dependence between the operations which stem from the action and its interiorization and this logic of the coordinations of actions: For example, the operation of adding two numbers $(2 + 3 = 5)$ proceeds from the action of reuniting objects and, if this symbolical reunion must be taxed, this is because the terms $2, 3, 5, =,$ and $+$ are signs and not things, but the addition bearing on these signs is, as reunion, a reunion as real as an addition bearing on objects.

Furthermore, it is worth insisting on the fact that operations, resulting from the interiorization of actions and from their coordinations, long remain relatively independent of language. Thus between the ages of seven and twelve, that is, before the construction of propositional or hypothetico-deductive operations which remain closely bound up with speech, we observe a long period characterized by concrete operations (categories, relations, and numbers) linked to the manipulation of objects themselves. These operations manifest themselves, among other ways, by the constitution of notions of preservation more general than that of the permanent object: in the proof of the balls of clay, for example, at about the age of seven to eight, conservation of substance, that of weight at about nine to ten, and that of volume at about eleven to twelve. Despite these chronological lags, the child, in order to justify these successive conservations, uses exactly the same arguments, expressing himself in rigorously identical verbal expressions such as, "It was only lengthened (the ball of clay into a sausage), nothing was removed or added, it's longer but thinner, etc." This clearly indicates that such notions do not depend only on language. It is a question, on the contrary, of a progressive structuration of the object according to its different qualities and as a function of the systems of active operations stemming from the actions as such exerted on the objects rather than the verbal formulation.

(b) The formation of thought as conceptual representation is assuredly correlative, in the child, with the acquisition of language, but in the first of these operations we cannot see a simple causal result of the second, for both are bound up with a still more general operation which is the constitution of the symbolical function.[1] Indeed, language appears at the same level of development as symbolical play, deferred imitation, and doubtless, mental image as interiorized imitation. The symbolical function itself in these various aspects is the differentiation of differentiated meanings and the capacity, due to these differentiated meanings, to evoke, the not actually perceived intended. These two characteristics oppose the verbal signs and the playful, gestural, or imagined symbols to sensori-

---

[1] One of the linguists invited to the Neuchâtel meeting of our Association remarked that it would be better to speak of "semiotic function" since it covers not only the use of symbols but also and above all that of "signs" (verbal, etc.) which are not symbols in the strict sense.

motor signs and signals nondifferentiated from their signified ones and thus incapable of serving to recall currently
nonperceptible objects or events. Transition between sensorimotor behavior and symbolical or representative behavior is probably assured by imitation (a proposition
common to Wallon's work as well as to our own) whose
deferred prolongation and interiorization assure their differentiation of significants and signified. This is especially
notable in a context of imitation acquired by language,
and this imitative factor seems to constitute an essential
auxiliary, for if the learning of language were due only to
conditioning, it would come much earlier. But since the
development of imitation itself is bound up with that of
intelligent behavior on the whole, we thus see that if it is
legitimate to regard language as playing a chief role in the
formation of thought, this is so to the extent that it constitutes one of the manifestations of the symbolical function,
the development of the function being in turn dominated
by intelligence in its total functioning.

(c) Once acquired, language is by no means sufficient
to assure the transmission of ready-made operatory schemes
which the child would thus receive from without by linguistic coercion. In this respect, a certain number of facts
can be stated. (1) Despite the classifications included in language, it is only on the level of concrete operations (seven
to eight years of age) that the child can master the handling
of inclusive definitions (by specific kind and difference:
Binet's and Simon's definition test) and of the classifications in general (Inhelder's and Piaget's third stage). (2)

Verbal expressions connoting the inclusion of a subcategory in a category, such as "Some of my flowers are yellow," are mastered only at the level where inclusion itself is assured, due to a play of additive and multiplicative operations of categories. (3) The practice of spoken numeration is not at all sufficient to assure the conservation of numerical sets nor that of equivalences by one-to-one correspondence, etc.

In short, adequate verbal transmission of information relative to the operatory structures is assimilated only on the levels where these structures are elaborated on the basis of actions themselves or operations as interiorized actions, and if language favors this interiorization, it neither creates nor transmits ready-made these structures by an exclusively linguistic means.

2. As for the problems relative to the necessary (although insufficient) role of language in the completion of operatory structures, these can first be studied with the greatest clarity at the level of formal or hypothetico-deductive operations, for these operations are no longer related to the objects themselves as concrete operations but to statements and hypotheses verbally expressed, etc. The propositional operations which are thus formed from the ages of eleven or twelve to fourteen or fifteen are obviously connected more closely to the exercise of verbal communication, and it is difficult to see how they develop or rather how they complete their development without the use of language.

(a) It is worth noting, however, that if the operations

take root "on this side" of language in the coordinations
of actions, they outdistance it "on the other side," in the
sense the propositional operatory structures constitute,
even if their elaboration is based on verbal behavior, relatively complex systems not included as systems in language
itself. These systems are, on the one hand, combinatory
(in opposition to the simple interlockings of concrete operations) and, on the other hand, a group of four transformations coordinating inversions and reciprocities—that is, the
two forms of reversibility previously separated into concrete groupings (classification, seriation, etc.). These two
correlative whole structures are seen in the subject's behavior by the construction of a set of new operatory
schemes (double systems of reference, proportions, proba-
bilities, combinatories, etc.) whose functional unity they
assure and whose relatively synchronic appearance they
explain. These large total structures outdistance the subject's language and could not even be formulated with the
sole aid of current language.

(b) The elaboration of such structures raises still
other problems and the question of knowing whether, in
this respect, language plays an authentically constitutive
role or only an indirect and auxiliary one, seems to us to
be reserved for the time when work stemming from various
trends of linguistic structuralism (Hjelmslev, Togeby, Har-
ris, and others) will have discovered sufficient connecting
points with the algebraic and logistic analysis of the mechanisms of thought.

(c) It already seems possible, on the other hand, to

foresee that, still at this level of formal or propositional operations, language acts less by transmission of ready-made structures than by a kind of education of thought or of reasoning due to the conditions of communication and to the precorrections of errors. "It would be possible," said the linguist Hjelmslev, "to reduce the system of formal logic and that of language to a common principle which could be called a *sublogical* system." L. Apostel has shown that this common system stems from the code theory and affirms the precorrection of errors which might occur between incoding and decoding. Thus it is in the direction of common functioning and of common probabilist source that we can conceive the formative action of language on operations, it then being understood that this formative action outdistances the language settings themselves and, on the basis of the coordination of the social actions, continues the equilibratory operation already active in the field of the coordination of actions in general.

At the close of these thoughts, we would like to stress that the research be continued in order to solve the preceding problems, to verify the proposed solutions, or to raise new questions.

A fruitful method, in this respect, consists of studying the effects of learning which bear on the verbal formulation of operations which have not yet been spontaneously acquired by children. The first research was done in this direction by A. Morf on children situated at an intermediary level between concrete and formal operations. Using some reasoning problems including implications, disjunctions, etc., Morf began by analyzing the children's spontaneous solutions, then he supplied them with verbal information by repeating the questions with additional precisions or by furnishing similar examples, etc. The results of these interventions were systematically negative except with children who had succeeded in solving one or the other of the questions by hypothetico-deductive means and who were then able to assimilate the significance of the additional information in the case of questions initially resulting in failure. Research of another type was undertaken by Inhelder and Bruner in recent experiments at Harvard University. Children uninformed of conservation in the example of liquids poured from one container to another were submitted to verbal learning based on expressions such as "glass A is both higher and thinner than glass B," etc. The questions were to analyze how the children learned to understand these relations and if they would modify their judgments on conservation (the experiment is done on the border of conservation tests but in the same context). The first facts gathered seem to show that: (a) the difficulties encountered in the progressive understanding of these verbal expressions are of the same order as the obstacles known on the field of the acquisition of conservation; (b) there are few connections between the two fields of verbal understanding and of concrete reasoning, as though it were a question at this level of two different projects.

A second instructive method is the analysis, currently

being carried out by J. Ajuriaguerra and B. Inhelder, of the relations between the linguistic level and the operatory level in the area of development troubles pointed out in one or the other of these two fields. We will say nothing of this, except to emphasize the interest of the paradoxical cases in which a strong linguistic backwardness is unaccompanied by trouble with intellectual operations themselves (a case in which one often finds the complete reciprocal: operatory backwardness without language troubles).

Finally the method of choice, in respect to the problems we have raised, is naturally that of the analysis of intellectual operations in deaf and dumb persons who possess the symbolical function without achieving an articulated language. The fine work done by P. Oléron, M. Vincent, and in our Center by F. Affolter has already shown that if these children show, according to tests, a varying backwardness in relation to those who hear, revealing among others things inferior mobility, they are nevertheless apt to master essential operations: classifications, seriations, and other order operations, perspectives (shadow tests), etc. A new problem, however, has recently been raised by Oléron as to an eventual considerable backwardness of these subjects in the acquisition of notions of conservation. However, the solution of such a problem is made quite difficult for method reasons, and we can ask ourselves if Oléron's results are not partially caused by his technique, all the more so as the means indicated for the norms do not wholly correspond to the tables recently established by various authors in several different countries. Affolter has turned again to the problem using other techniques and this already seems to indicate an earlier acquisition by the deaf and dumb, and the question remains open, like many others unfortunately, among those which we have raised.

7 *What Remains of the Gestalt Theory*

in contemporary psychology of intelligence and perception

Two principles of gestalt psychology remain fundamental in the fields which interest us in this study.

The first is that any operation stemming from perception or intelligence is characterized by a step toward equilibrium. This has been stated often in the field of affectivity where, according to Claparède, need reveals an imbalance and the satisfaction of a reequilibration. But it was probably with the theory of form that the notion of equilibrium acquired precise significance in the area of cognitive functions and inspired a series of decisive experiments: Hence the notion of field effects in the study of perceptions, and that of reequilibrations by successive levels in the study of intelligent acts, are notions which seem definitively acquired.

The second essential principle is that the forms of equilibrium forming at the close of these equilibration operations, consist of total structures characterized by organization laws stemming from the totality as such and not by association among previously isolated elements. The notion of totality, which became common as early as 1890 and the beginning of the twentieth century, also received from the gestalt theory a precise form whose significance is experimental.

We will fully retain these two central notions of gestalt psychology. Our goal, in this critical examination, is not to question them but, on the contrary, to take the analysis even further. We will retain in particular each of the organization laws (laws of "good form," etc.) revealed by gestalt studies on perception and almost every fact (with the exception of several exaggerations in respect to the so-called absence of development with age of certain mechanisms, in regard, for example, to the perceptive constants).

However, we will consider the forms of equilibrium or total structures described by the psychology of form as remaining incomplete: They are valid in certain fields, whereas in others, they seem to us to give way to other total structures which had not been foreseen by the initial form theory.

If we complete the gestalten by these other total structures which we will no longer call gestalten, it becomes necessary to make certain important changes to the theory, in view of the change of perspective due to additions.

The intelligence theory

Gestalt theoreticians began with perception or soon were dominated by perceptive models, even when those models were used to study intelligence.

Perceptions form nonadditive totalities and, consequently, essentially irreversible ones (in the logical sense of the word). As such, they correspond well to certain *physische Gestalten,* as Koehler showed, that is, to irreversible systems characterized by their displacements of equilibrium and in which each part is constantly subordinated to the whole according to a nonadditive mode of composition.

Hence the conception which Koehler gave us of gestalt and which we can define precisely by this essential character of nonadditivity. From such a viewpoint (and to remain with the physical gestalt so as not to prejudice the psychological statements which follow), the composition of forces in classical mechanics does not form a gestalt, whereas a soap bubble or the surface of undisturbed water does form a gestalt.

I am fully aware that for other authors the term *gestalt* tends to be applied to many other varieties of total structures. But if all is gestalt, I fear that this notion will lose its characteristic flavor. Let us remain therefore, for this discussion, with Koehler's definition which corresponds to the classic conception of the gestalt theoreticians.

When these theoreticians began to study intelligence, they used the same model as they used for perception. Conceiving the act of intelligence as a restructuring of the facts or as the passage of an inferior structure to a superior one, they conceived every structure, perceptive or intellectual, as obeying the same laws which they conceived as general.

It is on the first point that we find it difficult to follow the classic gestalt theory. Indeed, intelligence also obeys laws of equilibrium and laws of totality, but it is no longer a question of the same total structures as the perceptions: The total structures intervening in the field of logical and mathematical operations are, contrary to the perceptive structures, characterized by their reversibility (in the form of inversion or of reciprocity) and by their additive composition.

Take, for example, the series of whole numbers. Such a structure is a model of additive composition since $2 = 1 + 1$. And yet, it is really a question of a totality with its laws of organization, since a whole number does not exist independently of the series characterized by the operation $n + 1$ and since as totality this series presents structural laws which are those of the group (and of the "body," if we join in a single system the additive group and the multiplicative group). Here then is the case of an additive and reversible structure which, psychologically, presents every characteristic of an organized totality and which is not a gestalt in the precise sense of the term.

We can say the same of every logical structure whose development we studied step-by-step in the child and in the adolescent. Whether we consider classifications and seriations, correspondences, etc., in the field of group-ments of concrete operations (categories and relations) or the groups and networks of propositional operations, etc., everywhere we rediscover totalities of total structure characterized by laws of organization outdistancing the particular elements. And yet it is a question of reversible structures with additive composition: The initial intuitions (equilibrium and totality) of the gestalt theory are here deeply verified, and yet we find ourselves in the presence of totalities which are not gestalt.

Thus there exist two distinct forms of equilibrium at the core of cognitive operations (as elsewhere in the physical world):

1. The systems whose *conditions of equilibrium* are *permanent*. (For example: Logical operations and elementary mathematics which, once developed, are retained during a lifetime.) These systems are essentially *reversible,* which is not surprising since the equilibrium of a system is defined by the compensation of all its virtual transformations, that is, precisely by reversibility.

It is these systems which appear to us to be characterized by logical thought, beginning at the level of concrete operations (as early as the age of seven or eight) and which thus constitute the forms of equilibrium to which every development of intelligence tends (indicatory manifestations are observed already from the time of the infant's or chimpanzee's sensorimotor intelligence with the constitution of the permanent object and of the practical group of displacements).

2. The systems with *forms of momentary equilibrium,* hence characterized by their *displacements of equilibrium* and which, consequently, are only *semireversible* or more or less *irreversible.* In such systems, the logico-mathematical operations are not yet possible, but they are replaced (and prepared) by semireversible regulations or mechanisms.[1] Such are the systems which we encounter in the fields of perception and of preoperatory intelligence, where the gestalts of nonadditive composition play a full role.

But, if this is so, we would no longer be able to consider intelligence a continuation of the perceptive structures. The reversible mobility of the operatory intelligence does not result from a mere softening of the nonadditive gestalts but, in the majority of cases, requires a kind of melting or dissolution of these elementary configurations. The great difference between the child of four to six and the one of eight to ten is that the first, for his reasoning, turns to perceptive *configurations* or gestalts, whereas the second reasons on the *transformations* which lead from one configuration to another. To subordinate the configurations to the transformations is not only to free the first but to involve them in a movement which transforms them by making them lose their nonadditive and irreversible characteristics.

---

1 Which recalls in physics the operation of moderation, as in Le Châtelier's principle.

Wertheimer, in his celebrated analyses, sought to reduce to gestalts the inclusions of categories intervening in
syllogism and to reduce the conclusion of syllogism to an
Umzentrierung (centration) which decenters a class (or an
individual) from another to recenter it in a third. This
interpretation, however, would be insufficient for us: Logical truth is not due to one *configuration (A* included in *B*
or *B* in *C)* or another *(A* included in *C)* but to the very
system of the *transformations* $(A + A' = B; \ B + B' = C;$
$C - B' = B; \ B - A' = A,$ etc.). These transformations
no longer form a gestalt comparable to a figure, since in
essence they are reversible (+ and —) and of additive composition. Wertheimer was probably committed to this direction: The deeply moving posthumous volume which
his friends published [2] constantly mentions grouping and
operations. But so long as the duality of the nonadditive
and reversible structures is not recognized as fundamental,
there remains a somewhat disturbing ambiguity in these
attempts of unification of logical forms and of gestalts as
such.

If intelligence forms an operatory activity leading to
the formation of reversible structures instead of consisting
simply of a restructuration according to laws identical
with those of the perceptive structures, it then becomes
necessary to restore to intelligence a certain number of
characteristics which appear to have been somewhat neglected by the gestalt description.

---

[2] M. Wertheimer, *Productive Thinking* (New York and London: Harper
& Brothers, 1945).

The first of these characteristics is the constructive activity of intelligence. In "operating" on the objects, the subject alone, by his action, elaborates structures and is not simply the theatre of a restructuration or of a reequilibration acting according to the laws of gestalt physics. Otherwise (and this is the impression we often have when reading certain gestalt works) the simultaneous restructuration of objects and subject within the "field" which encompasses them both, leads us back to a kind of empiricism by suppression of the constructive role of operations, with the sole difference that the total structures are substituted for the associations of the former associationist empiricism.

Nor could the reversible structures of intelligence be assimilated to innate forms. Gestaltists with maturationist tendencies often present the gestalt laws of organization as structuration conditions independent of any experience, in other words, innate or *a priori*. But the logical necessity characteristic of the structures of intelligence is never given *prior* to the experience. On the contrary, it forms itself *at the end* of an evolutionary operation which partly depends on practice and experience, being a *final* necessity, as in the case of the forms of equilibrium (independently, this goes without saying, of all finalism) and not initial, as in the case of innate or hereditary systems. Thus it is quite distinct from a "pregnance" inasmuch as the latter would result from an inherited mechanism. (It is distinct moreover from any perceptive "pregnance" for reasons we have just seen.)

Actually, the subject depends neither solely on external objects (configuration of the field) nor solely on its innate mechanism, but reveals an activity which is bound up with its own history. The structures of intelligence are not extemporaneous gestalts but schemes which derive one from the other by progressive relation during a continuous construction. Associationist empiricism would consider these schemes the mere result of previous experience. Duncker rightly replied that the subject turns to the past for what he needs as a result of the present situation. In fact, the present structure is a scheme which proceeds from previous schemes but which reacts on them by integrating with them. The integrative conception applied to the historical and constructive activity of intelligence thus outdistances simultaneously the two contrary theses by regranting a legitimate role to experience but without reducing to the latter the forms of equilibrium themselves.

Finally it is indispensable to restore to the act of intelligence the element of control or of correction (often revealing itself by groping itself) which one often has a tendency to forget.

The theory of perception

On this elementary ground, gestalt psychology was completely right to insist on the existence of structures irreversible to nonadditive composition (gestalt in the strict sense) and on their laws of organization.

But, even in this field particularly favorable to its theses, we do not believe that the gestalt theory corresponds to all the perceptive facts. It is, in fact, very probable that the perceptive mechanisms are not distributed on a single level: There are field effects or immediate interactions among elements perceived simultaneously in their mutual relations, and it is on this fundamental basis that the gestalt description retains its whole value (though moreover it can be outdistanced in the direction of the analysis of relations but without this contradicting the gestalt laws). On the other hand, beyond the field effects, there exists a set of perceptive activities placing in relation, at ever greater distances in space and time, the elementary effects, and these perceptive activities extend more and more beyond the gestalt setting because they orient in the direction of intelligent reversibility.

Révész,[3] in an interesting article, believes that the gestalt theory should be revised as it concerns tactile perceptions, for these are not simultaneous and require a continual positioning among successive facts. Along with B. Inhelder,[4] we have made similar remarks in regard to the development of stereognosia in the child aged four to seven or eight: Whereas these children recognize the forms poorly because they remain more passive, we see a development of an increasingly systematic exploratory activity which, for tactile forms measuring a few centimeters, alone allows children to discern the Euclidian spatial characteris-

---

3 G. Révész, *Zur Revision der Gestaltpsychologie, Revue suisse de Psychologie*, vol. XII (1953), pp. 89–110.

4 Piaget and Inhelder, *La représentation de l'espace chez l'enfant*, chapter II.

tics (in opposition to the elementary topological characteristics).

This exploratory activity, however, is far from being special to tactile perception. In the visual field, the gestaltists were forced to admit that it also modifies structurations, and they then spoke of "analytical attitude" (but the analysis is not a mere attitude but an activity itself). The question then is to establish if such an activity still stems simply from the gestalt laws or if it is removed from them in varying degrees.

Here is an example. In turning again to an idea developed in a posthumous article by the late Rubin we studied with Maire and Privat the resistance of good forms between the ages of five or six and the adult age, when combining a square with the Müller-Lyer illusion (by adding external feathers to the upper side of the square and internal ones to the lower side). Instead of presenting the same degree of "pregnance" at any age, as would have conformed to the theory, it was found that the square form is about three times less resistant (relative to the respective values of the Müller-Lyer illusion) among children than adults. Thus, in the case of good form, two distinct effects intervene: a field effect which exists at every age but which gives rise only to a form remaining somewhat elastic, and an effect of comparison (analysis) between the sides or between the angles. This last effect, which becomes increasingly important with age, then gives rise to a perceptive scheme, transposable by assimilatory recognition and generalization and not simply by automatic and independent restructurations; this scheme is then far more resistant than the primary good form due to the field effects alone.

In a general manner, there thus exists perceptive activities outdistancing field effects and which correspond to what is ordinarily called analysis or exploration. They consist of transports in space and time, of double transports or comparison (transport of one of the terms compared with the second and reciprocally), of active transpositions (or transport of a complex of relations with recognition and generalization), of anticipations or *Einstellungen*, in relation with more and more distant reference elements, etc. This perceptive activity depends on the motricity, on the postural system (compare H. Werner's sensoritonic theory), and finds itself in closer and closer liaison with intelligence by the intermediary of the sensorimotor schemes.

The distinction between field effects and those of perceptive activity corresponds to a genetic criterion: The first somewhat diminish in importance with age, while the second increase in value during development. It is in this sense that the genetic study of perceptions, which we have been conducting with Lambercier for the past twelve years, appears to us to furnish an additional dimension to the analysis of the perceptive mechanisms, whereas research of the constant laws of organization during development has become somewhat too exclusive among the majority of gestaltists (with certain notable exceptions, for example, in Meili's work) and has prevented sufficient perception of the multiplicity of planes on which perception is organized.

If we return to field effects or primary effects, we can ask ourselves whether this change of perspective is not natural, not to upset the gestalt conceptions of the laws of organization and good forms, but to allow for greater depth in the sense of a more relativist theory and above all of a more quantitative one.

Indeed, it is surprising that after furnishing an excellent qualitative description of field effects, the gestalt theory has not elaborated quantitative laws of perceptive distortions and of good forms. It is likewise surprising that all the characteristics attributed by the theory to the perceptive good form (simplicity, regularity, symmetry, resemblance, proximity, etc.) find themselves constituting moreover qualities essential to the logico-mathematical structures (except proximity, which intervenes as closeness in topological structures). The gestalt description has thus remained too overall to achieve quantitatively (and even partly qualitatively) what differentiates primary structures (field perceptive effects) from structures of intelligence. But why?

The reason, it appears, is that the notion of totality is a notion whose fascinations are dangerous. Excellent as a descriptive notion, it first appears to constitute an explicative notion, whereas it alone never explains anything. To say and to repeat constantly that "the parts are distorted by the whole" is not an explanation: It is a program and a good program of future explanations, but the real explanation only begins when one succeeds in placing the part in liaison with the set of the other parts according to a system of relations themselves. The former associationism first placed the isolated elements and constructed the whole by associations among them. The merit of the gestalt theory was to show that there is totality at once, but it placed too much weight on the whole as a type of cause which acted on the parts. A third conception is possible which retains moreover the essential of the second, namely, that the whole consists, from the beginning, of a system of relations (not of elements but of relations whose elements are at once bound up); these relations can then be studied and formulated in their very interdependence, allowing the elaboration of quantitative laws.

We have thus tried to free the law which makes it possible to determine the *maxima* and the *minima* in the cases of geometrical deceptions, a factor of which is varied: a rectangle, for example, whose deception is measured on one of the unchanging sides, while the other varies, or the Oppel delusion with measure of the constant length of a divided line, while the number of divisions varies, etc.

This law [5] appears to explain itself by actions easily inscribed in the setting of perceptive activity previously described while maintaining the originality of the field effects.

To interpret the field actions stemming from such a law, one merely has to turn to a mechanism whose effective

---

[5] See *Proceedings and Papers of the Thirteenth International Congress of Psychology at Stockholm,* 1951, p. 197. The law has the following form:

$$P = \frac{(L_1 - L_2) \times n \ (L_2/L_{max})}{S}$$ where L_1 *is the largest of the compared* lengths, L_2 the smallest, L_{max} the total length of the figure, n the number of the Ls, and S the surface. P is the relative value of the illusion.

influence we have tried to control. Any element in the center of a glance (in the case of visual perception, but the phenomenon is found again in other fields) is by this very fact overestimated, whereas the elements noncentered are devalued in relation to it, the overestimations or under-estimations being proportional to the sizes of the elements considered. By using such an effect, we can then simultaneously explain the actions of contrast between two unequal sizes (when this inequality outdistances the value of the overestimations by centration) and the actions of equalization between two neighboring sizes (when their inequality is inferior to this same value). That is why an essential factor of geometrical deceptions is the difference between the two compared principal sizes $(L_1 - L_2)$.

But, in addition to these actions of centration which are therefore fundamental in field effects, decentration actions also intervene when several successive centrations are placed in relation and which result in a set of relative compensations (hence the relations L_2/L_{max} and $1/S$ which also intervene in the quantitative expression of the deceptions). This decentration already constitutes a beginning of perceptive activity, and that is why we just stated that the study of relations intervening in field effects is inscribed in the setting of the perceptive activity in general the moment one proceeds by detailed and quantitative composition of these relations.

However, our goal here is not to explain the result of our research on these complex questions but merely to show that, while being inspired by notions of equilibrium and totality characteristic of gestalt psychology, and while retaining on perceptive grounds the notion of irreversible totalities with nonadditive composition (gestalt proper), we can take the analysis even further by placing ourselves at a viewpoint which is both more relativist and more quantitative. It is worth adding that, on these grounds, any quantitative analysis leads to a mode of probabilist composition of perceptive structures, this probabilist character of perception opposed to the character of intrinsic necessity peculiar to the logico-mathematical structures doubtless being what precisely accounts for the opposition between the irreversible or nonadditive compositions and the reversible compositions.

In conclusion, much remains to be retained of the gestalt theory in present studies of intelligence and perception. We ourselves believe that we retain the essential with the notions of equilibrium and of totality (or of organized structure sets).

However, it is necessary to complete the gestalt viewpoint by recalling still other structures, as we did concerning reversible structures of intelligence. But, proceeding in this manner, we are not far from the basic hypotheses which inspired early gestalt work; in a sense, we are more faithful to these authors than they were to themselves. In fact, to turn to the notion of equilibrium is not only to commit ourselves to use every form of equilibrium, and not exclusively these particular forms constituted by gestalt in the strict sense, but also, and above all, to commit ourselves to the full use of the notion of reversibility, for equilibrium is precisely defined by reversibility! Whether such experiments are considered contrary to the gestalt theory, as is the opinion of certain orthodox partisans of the School, or whether they are called Neogestaltists, which our friend Meili did one day, all this is of no importance. In conclusion, we would like to express our debt to a doctrine which has considerably influenced contemporary psychology and to which independent researchers are as indebted as are others.

8 The Necessity and Significance of Comparative Research in Genetic Psychology

Genetic psychology is the study of the development of mental functions insofar as this development can offer an explanation or at least a complement of information concerning their mechanisms at the finished state. In other words, genetic psychology consists of using child psychology to find the solution to general psychological problems.

From this viewpoint, child psychology forms an irreplaceable instrument of psychological investigation. We are becoming now more and more aware of this, yet less aware of the fact that its role could become almost as important in sociology. Auguste Comte rightly claimed that one of the most important phenomena of human society is the formative action of each generation on the following one, and Durkheim arrived at the collective origin of moral feelings, legal norms, and logic itself. But there is only one experimental method to verify such hypotheses, namely, the study of the individual's progressive socialization, that is, the analysis of his development in terms of the particular or general social influences which he undergoes during formation.

Any comparative research dealing with different civilizations and social milieus poses from the very outset the problem of the delimitation of factors peculiar to the individual's spontaneous and inner development and of the collective or specific cultural factors of the surrounding society considered. This delimitation, which we cannot ignore, can lead to unexpected results. In the field of affective psychology, for example, the early Freudian doctrines furnished the model of an endogenous individual development, so endogenous that the different proposed stages, especially that of the so-called Oedipal reactions, were presented as due essentially to successive manifestations of one and the same "instinct," that is, inner tendencies which owed nothing to society as such. On the other hand, we are well aware that a whole group of contemporary culturalist psychoanalysts (including Fromm, Horney, Kardiner, and Glover, along with Ruth Benedict and Margaret Mead) now support the hypothesis of a close dependence of the various Freudian complexes, especially Oedipal tendencies, to the surrounding social milieu.

Development factors

In the field of cognitive functions, the only one we will be concerned with in what follows, the principal advantage of comparative research is to allow for the dissociation of individual and collective development factors. Here again, however, it is worthwhile to introduce first of all a few necessary distinctions regarding the factors to be considered.

1. *Biological factors.* First, there are biological factors linked to the epigenetic system (interactions of the genome and of the physical milieu during growth) which are revealed especially by the maturation of the nervous system. These factors, which doubtless owe nothing to society, have a role still scarcely known, but their importance probably remains equally decisive in the development of cognitive functions, and we should therefore keep in mind the possibility of this influence. In particular, the development of an epigenotype implies, from the biological viewpoint, the intervention of stages which show *sequential* character (each being necessary to the following one in a constant order), *creodes* (canalizations or passages necessary to the development of each special section of the whole), and a *homeorphesis* (kinetic equilibrium in the sense that a deviation in relation to creodes is more or less compensated with a tendency to return to the normal path). These are characteristics which until now we thought we could find in the development of operations and of the logico-mathematical structures of intelligence. If the hypothesis were true, this would naturally suppose a certain constancy or

uniformity of development regardless of the social milieu within which individuals are formed. Inversions in the series of stages or important modifications in their characteristics from one milieu to another would prove, on the contrary, that these basic biological factors do not intervene in the individuals' cognitive evolution. Here then is the first basic problem whose solution requires extensive comparative research.

2. *Equilibration factors of actions.* The study of the development of intellectual operations in numerous culturally developed countries where the study of our stages was undertaken, immediately shows that psychological factors are far from being the only ones at work. If a continuous action of the internal maturation of the organism and of the nervous system alone intervened, the stages would not only be sequential but also linked to relatively constant chronological dates, as is the case of coordination of vision and prehension at about the age of four to five months, the appearance of puberty, etc. According to individuals and their family, school, and social milieu, we find generally in children of the same city often considerable advancement or retardation, not inconsistent with the order of succession which remains constant but which reveals that other factors are added to the epigenetic mechanisms.

A second group of factors should therefore be introduced, holding in reserve its possible connections with social life which, in principle, again emanate from activities peculiar to general behavior in its psychobiological as well as in its collective aspects. These are equilibration factors taken in the sense of self-regulation and hence in a sense closer to homeostasis than to homeorhesia. In fact, individual development is the function of multiple activities in their aspects of exercise, experience, or action on the milieu, etc. Hence individual development constantly intervenes between these actions of particular or increasingly general coordinations. This general coordination of actions supposes multiple systems of self-regulation or equilibration, which will depend on circumstances as much as on epigenetic potentialities. These very operations of intelligence can be considered the superior forms of these regulations, revealing both the importance of the equilibration factor and its relative independence in relation to biological preformations.

Here again, however, if the factors of equilibration can be conceived as very general and relatively independent of particular social milieus, the hypothesis demands comparative verification. Such processes of equilibration are especially noted in the construction of conservation notions, the stages of which reveal in our milieu not only a sequential series but also the elaboration of compensation systems whose intrinsic characteristics are very typical of these regulations by successive layers. But are the particular stages found everywhere? If so, we would still not have verification of the hypothesis but at least a somewhat favorable indication. If not, on the contrary, it would be the mark of cultural and of particular and variable educative influences.

3. *Social factors of interindividual coordination.* Coming now to social factors, it is worthwhile to introduce an essential distinction, equally important, in the psychobiological field, namely, that of epigenetic potentialities and effective regulations or equilibrations that become evident or are formed during activities peculiar to behavior. This distinction is that of interactions or general social (or interindividual) coordinations that are common to all societies, and transmissions or cultural and especially educative formations which vary from one society to another or from one restricted social milieu to another.

Whether we study children in Geneva, Paris, New York, or Moscow, in the Iranian mountains, in the heart of Africa, or on a Pacific island, everywhere we observe certain social conduct of exchange between children or between children and adults, which takes effect by their very functioning, independent of the contents of educative transmissions. In every milieu, individuals gather information, collaborate, discuss, oppose one another, etc., and this constant interindividual exchange intervenes during the whole development according to a socialization operation which concerns not only the social life of children among themselves but also the relationship with their elders or with adults of any age. Just as Durkheim refers to general social mechanisms, claiming that "civilization lies beneath civilization," so in order to treat the relation between cognitive functions and social functions, we must begin by opposing the general coordinations of collective actions to the particular cultural transmissions which crystallize in a different manner in each society. Thus in the event that we rediscover our stages and results in every society studied, this will in no way prove that these convergent developments are strictly individual. As it is evident everywhere that the child benefits from the youngest age from social contacts, this would show also that there exist certain common operations of socialization which interfere with the operations of equilibration previously studied.

These interferences are so probable, and probably so slight, that we can at once assume what would be confirmed or revealed by future comparative studies, namely, that, at least in the field of cognitive functions, the general coordination of actions, whose progressive equilibration appears constitutive of the formation of logical or logico-mathematical operations, might involve not only collective or interindividual actions but individual actions as well. In other words, whether it involves actions done individually or those done in common with exchange, collaboration, opposition, etc., we find the same laws of coordination and regulation which result in the same final structures of operations or cooperations as co-operations. We could thus consider logic, as the final form of equilibrations, simultaneously individual and social, individual insofar as it is general or common to every individual, and likewise social insofar as it is general or common to every society.

4. *Factors of educative and cultural transmission*

On the other hand, in addition to this functional core in part synchronic yet susceptible to construction and evolution, characteristic of interindividual exchanges, we must naturally consider the chief diachronic factor consisting of cultural traditions and educative transmissions which vary from one society to another. When speaking of social factors, we generally think of these differential social pressures, and it goes without saying that, insofar as cognitive operations can vary from one society to another, we ought to consider this group of factors, which is distinct from the preceding one, beginning with various languages capable of exerting more or less strong action, if not on the operations themselves, at least in conceptualization detail (contents of classification, of relations, etc.).

Comparative research in the field of cognitive operations

Once we have established the four kinds of factors in this classification, according to the general types of relations between the individual and the social milieu, let us now try to establish the essential unity which comparative research would offer concerning our knowledge of cognitive operations. The central problem in this respect is that of the nature of intellectual operations, especially of logico-mathematical structures. A certain number of hypotheses are possible which, among other things, correspond to the four preceding distinct factors, with additional subdivisions.

Biological factors and factors of action coordination. The first interpretation would consist of considering them if not innate, at least resulting exclusively from biological factors of an epigenetic nature (maturation, etc.). It was in this direction that K. Lorenz turned. Lorenz, one of the founders of contemporary ethology, believes in *a priori* knowledge and interprets it on the mode of instincts.

From the viewpoint of the comparative facts that we have been able to gather and can still gather, two questions arise: Do we always find the same stages of development, naturally taking into account corrections and eventual improvement for the known tables? Will we always find them at the same average ages? To reply to these two kinds of questions, it is in addition useful, and even almost necessary, to have reference elements, by comparing the evolution of reactions with operatory tests (conservations, classifications and inclusions, seriations, numerical correspondences, etc.) to the evolution with the age of reaction to tests of simple intellectual performance, similar to those generally used to determine IQ.

Comparative research has only just begun, and it would be rash to draw conclusions so early, considering the material we should have and the great linguistic and other difficulties, not to mention the long initiation necessary to master study methods, these being all the more difficult to use since they bear more on operatory functioning. But the early work offers a glimpse of certain results which at least indicate what might be an interpretation line if they prove generalizable.

In Iran, for example, Mohseni, in 1966, questioned both children schooled in the city of Tehran and young country illiterates by means of conservation tests on the one hand and performance tests (Porteus, graphic tests, etc.) on the other. The three principal results obtained from children aged five to ten are the following. (a) In general, the same stages in city and country, in Iran and in Geneva, etc., were found (series of conservations of substance, weight, volume, and so forth). (b) There was a systematic difference of two to three years for operatory tests between villagers and city dwellers but at about the same ages in Tehran and in Europe. (c) The delay was considerable at the age of four and especially five for performance tests between villagers and city dwellers [1] to the point where the former appeared mentally defective without the operatory tests.

Assuming such results were found elsewhere, we would be led to the following hypotheses.

(a) A more general verification of constancy in the order of stages would tend to show their sequential characteristic in the sense indicated above. Until now this constant order seems to have been confirmed—in Hong Kong according to Goodnow (1962), in Aden according to Hyde (1959), in Martinique according to Boisclair, in South Africa according to Price-Williams (1961)—but it goes without saying that other facts are still necessary. Insofar as we could continue to mention sequential order, there would be an analogy here with the epigenetic development in the Waddington sense and, consequently, a

---

[1] Children schooled in Tehran are one to two years behind European and American children.

certain intervention probability of factor 1, specified above. But how far? In order to recommend with certainty the biological factors of maturation, we would have to be able to state the existence not only of a sequential order of stages but also of certain average dates, chronologically set, of appearance. Mohseni's results show, on the contrary, a systematic retardation of country children over city children which indicates, of course, the intervention of factors other than those of maturation.

On the other hand, in the field of representation and thought, we could perhaps find the same important date everywhere, that of the construction of the semiotic or symbolical function, which appears in our milieus between about the ages of one and two: formation of symbolical play, of mental pictures, and so forth, and above all language development. The principal factor which makes this semiotic function possible seems to be the interiorization of imitation. On the sensorimotor level, this interiorization already constitutes a kind of representation in action, as a driving copy of a model, in such a manner that its extensions, first in deferred imitation, then in interiorized imitation, allow for the formation of representation in pictures, and so forth. But these operations of deferred reactions, then of interiorization, naturally suppose certain neurological conditions, for example, a halt at the level of certain relays in the realization of schemes of action without complete effectuation. A comparative study of sensorimotor forms of imitation and dates of appearance of semiotic function based on deferred imitation, would perhaps show certain chronological regularities, not only in the sequential order of the stages but also in the more or less set dates of formation. In this case, we would come closer to the possible factors of maturation which are relative to the epigenetic system (intervention of language centers, and so forth).

(b) The second net result of Mohseni's research is the rather general retardation of country children over those in Tehran insofar as operatory tests (conservations) and performance tests are concerned. This difference thus proves with certainty the intervention of factors distinct from those of simple biological maturation. But here we can hesitate among the three groups of factors mentioned above (factors 2, 3, and 4), that is, factors of activity and equilibration of factors, factors of general interindividual interaction, and those of educative and cultural transmission. In fact, each of these factors can intervene. Insofar as factor 2 is concerned, the author noted the surprising deficiency of activity among the country children who in general are not only without schools but also without any toys other than pebbles and bits of wood and show rather constant apathy and passivity. We find ourselves therefore in the presence of both a weak development of coordinations of individual actions (factor 2), and interindividual ones (factor 3), and educative transmissions which are reduced since the children are illiterate (factor 4), implying a convergence of these three groups of factors combined. How then to distinguish them?

(c) On this point the third result obtained by Mohseni

is instructive. Despite the deplorable situation of country children, their reactions to operatory tests proved superior to their results on performance tests. Where we might consider them mentally defective or even imbeciles solely on the basis of intellectual performance tests, they are but two to three years behind Tehran children in conservation tests. Here again, it goes without saying that we cannot take the chance of generalizing without having numerous facts from milieus quite different. To show the interest of the problem and the many distinct situations which remain to be studied, mention should be made that Boisclair, along with Laurendeau and Pinard, began in Martinique to study a group of school children who were anything but illiterate, since they followed the French primary-school teaching program. Nevertheless, they showed a retardation of about four years in the principal operatory tests. In this case, retardation seems to have been due to the general characteristics of social interactions (factor 3 linked with 2) rather than to a deficiency in educative transmissions (factor 4).

In the Iranian case, the interesting advance of successful conservation tests, an indication of operatory mechanisms over performances used elsewhere, seems to indicate a dual nature between the rather general coordination necessary to the functioning of intelligence and the more special acquisitions relative to particular problems. If these results were increased, it might lead to separate factors 2 and 3 considered together (general coordination of action, whether individual or interindividual) from factor 4 of transmission and education. In other words, operatory tests would give rise to improved results because they are linked with the coordination necessary to any intelligence as products of progressive equilibration and not as previous biological conditions, whereas performances would undergo retardation as a result of more particular cultural factors and, in some cases, especially deficient ones.

Such are the general possibilities for exploration which might be offered by comparative facts similar to but in greater quantity than those gathered by Mohseni. But these are only general outlines, and we must now examine in greater detail the role of social factors stemming from groups 3 and 4.

Social factors of educative transmission. If operatory structures could not be explained, in accordance with the hypothesis we have developed, by the most general coordination laws of action, we would have to consider more restricted factors, of which the two principal ones might be, for example, an adult educative action similar to those which engender moral requirements, and language itself, as crystallization of syntax and semantics which, in their general forms, include logic.

(a) The hypothesis of the formative action of adult education certainly has some truth, for even in the perspective of general coordinations of action, material or interiorized in operation, the adult, more advanced than the child, can aid the child and increase his development during the course of family or school educative operations. But the question is to know whether this factor has an exclusive role. This was Durkheim's idea, for whom logic, like morals and law, stems from the total structure of society and imposes itself on the individual, due to social constraints and, above all, to educative ones. This is somewhat Bruner's idea also (1964): Turning to less scholarly educative processes and coming closer to the American models of learning, he believes that one can learn anything at any age, by going about it in an appropriate manner.

So far as Durkheim's perspective is concerned (but not that of Bruner, which depends on laboratory verifications [2] more than on comparative studies), facts, like those observed in Martinique by Canadian psychologists, seem to indicate that ordinary schooling with a French program to facilitate comparison is not enough to assure normal development of operatory structures, since in this case there are three or four years of retardation in comparison to children of other cultural milieus. But, here again, we must naturally not come to hasty conclusions, for there remains above all the task of dissociating family and school influences. All we are stating therefore is simply that the comparative method is, on this point as on others, apt to furnish solutions we have been looking for.

(b) As for the great problem of language in its interactions with operatory development, we are beginning to get a clear idea of it, following Sinclair's research on the child's linguistic development and Inhelder's and Sinclair's research on the role of language in learning experiments of the operatory structure.

---

[2] These verifications were undertaken in Geneva by Inhelder and Bovet and are rather far from verifying Bruner's hypotheses.

Without going into detail about methods and results, we will limit ourselves to emphasizing the perspective offered by Sinclair's research from the comparative viewpoint. Let us recall, for example, the experiment made on two groups of children, the older having clear conservation structures (with explicit arguments) and the younger group at a level equally unequivocal about nonconservation. The children of these two groups were asked to describe not the material used for these two previous determinations but certain objects attributed to personages represented by dolls (a short thick pencil, another long and thin; several marbles, a small number of larger ones). The language used in both groups, it was noticed in a very significant manner, differed according to the comparative expressions used. Whereas subjects without conservation used above all what the linguist Bull called "scalars" ("large" and "small," "many" or "few," and so forth), subjects of the operatory level used "vectors" ("more" and "less," and so forth). In addition, the structure of the expressions differed according to binary modes ("this one is longer and thinner") or quaternary modes ("here it is thick and the other is thin; here it is long and the other is short," and so forth). Here there is a strict correlation between operativity and language, but in which sense? Learning experiments, which do not concern us directly here, show that in training nonoperatory subjects to use their elders' expressions, only a slight operatory progress is achieved (one case out of ten). There remains, moreover, the question of establishing whether it is a question of language as such or of the influence of analysis exercises that learning involves, and whether certain progress would not have been accomplished without this learning by development of schemes as a function of various activities. It appears, therefore, that it is operativity that leads to structure language, by choice within pre-existing language naturally, rather than the contrary.

We at once see the great interest in increasing experiments of this kind as a function of various languages. Sinclair found the same results in French and English, but there remain other languages all quite different. In Turkish, for example, there is but a single vector which corresponds to the French *encore* (again). To say *more* we would have to say *again a great deal,* and to say *less, again little.* It goes without saying that many other combinations are to be found in other languages. In this case, it would be of great interest to study the time allowed to develop operatory structures as a function of the subject's language and to resume Sinclair's experiments on children of different levels. To suppose that evolution of thought structures remains the same despite linguistic variations would be a fact of some importance, which would argue in favor of progressive and autonomous equilibration factors. Supposing on the other hand, that there are operatory modifications according to linguistic milieus, the sense of these dependencies remains to be studied according to the experimental model suggested by Sinclair.

Conclusion

In short, the psychology which we are developing in our milieus, characterized by a certain culture, a certain language, and so forth, remains essentially conjectural until we have furnished comparative material for control purposes. And, in respect to cognitive functions, our comparative research concerns not only the child but also development in its entirety, including adult final stages. When Lévy-Bruhl raised the problem of "prelogic" peculiar to "primitive mentality," he doubtless exaggerated the oppositions, and likewise his posthumous retraction perhaps exaggerated in an opposite sense structural generalities. But a series of questions still remains unsolved, it seems to us, by the fine work accomplished by Claude Lévi-Strauss. For example, what is the adult operator level in tribal organization in respect to the technical intelligence entirely neglected by Lévi-Bruhl, verbal intelligence, the solution of elementary logico-mathematical problems, and so forth? Obviously, it is in knowing the solution, the development of adults themselves, that genetic data relative to lower age levels would acquire its full significance. In particular, it is quite possible, and this is the impression we have from known ethnological work, that in many societies adult thought does not go beyond the level of concrete operations, and therefore does not reach that of propositional operations, which develops between the ages of twelve and fifteen in our milieus. Thus it would be of great interest to know if the early stages develop more slowly in children of such societies, or if the degree of equilibrium which will not be exceeded is achieved, as with us, at about the ages of seven or eight or with merely with slight retardation.

9 *Life and Thought*

*From the viewpoint of experimental
psychology and of genetic epistemology*

1. A. Lalande, basing himself on arguments which elsewhere retain their entire interest, assigned to the evolution
of logical thought a direction contrary to that of fundamental evolution. Bergson has taken up this opposition.

The Spencerian hypothesis of a continuity between
the evolution of life and that of intelligence, remains nevertheless the most plausible, on the condition, naturally, of
being reworked as biological and psychological contributions are made.

In this respect, a series of works could be discussed.
We will limit ourselves to recalling the fine book by Ruy-
ssen, *L'évolution psychologique du jugement.*

2. We will restrict ourselves, by method, to using only statements and interpretations already developed in the field of scientific biology and psychology, trying to resist speculative temptations. Indeed, we believe that no philosophical psychology exists but only an experimental psychology and a philosophy of psychology in the sense of an epistemology of psychological knowledge. A philosophical psychology proposing to furnish corrections or additions to the results of experimental biology and psychology would appear to stem from the same kind of inspiration as nineteenth-century *Naturphilosophie* and be doomed to the same destiny.

3. Three kinds of problems now studied by psychologists are rather close to the central question of relation between life and thought: that of functional interactions between maturation of the nervous system and the milieu (experience), that of the structures in general and of their psychobiological significance, and finally (apropos notably of structures), that of psychophysiological parallelism or of isomorphism.

Functional interactions between the organism and the milieu

4. Whether we study the evolution of a perceptive mechanism (for example, the perseverance of size in depth, or the relation to good forms, etc.), the acquisition of a scheme of sensorimotor intelligence (for example, the scheme of the permanent object or the coordination of

movements according to a group of displacements, etc.), or even the important stages of development of child intelligence (for example, the set of transformations marking at about the age of seven, the transition of the preoperatory representations to the early systems of logico-mathematical operations), we constantly run up against the same problem which reappears in many forms and has failed to offer every researcher satisfactory solutions. That is the problem of the respective influence of hereditary systems (mechanisms present from the time of birth or subordinated to a progressive internal maturation) and of experience acquired or of exercise.

5. For certain authors, the cognitive mechanisms are acquired as a function of experience in every or in certain fields (see Pieron's radical position on the perceptive ground). For others, innateness plays an important and assignable role in the perceptive field (see Michotte's position elsewhere recently summarized), or in the sensorimotor systems (see Wallon, Bergeron, and others, before the renewal of Pavlovian reflexology). Most of the authors acknowledge two kinds of factors but refuse to assign an exact role in general or even in any particular area.

6. It is clear that such a psychological problem forms a special case of the general biological question of relations among the genotypical characteristics and the phenotypical characteristics. In this respect, we could attempt (we have insisted on it elsewhere) a kind of parallelism between the biological solutions of the problem of variation or evolution, the psychological solutions to the prob-

lem of intelligence, and the great epistemological problems themselves.

7. The discussion will probably remain in doubt so
long as, on the organic level as such, there remain the
mechanisms of growth (ontogenesis) in their relation with
heredity and above all the phylogenetic relation between
heredity and milieu. The absence of a solid third position
between mutationism and the hypothesis of acquired heredity seriously hinders not only the psychological explanation but the biological explanation as well.

8. But because of the lack of causal or structural analysis, it is interesting to free the functional analogies. It
should first be noted that, in biology, a genotypical characteristic in its isolated state is never known, for even in
the laboratory it is always associated with a phenotype
linked to the studied milieu. Thus genotype and phenotype do not form an antithesis on the same level, for the
genotype is the set of characteristics common to every phenotypical arrangement compatible with a determined pure
race, including the capacity to produce these arrangements.
In other words, we must consider not only actualized characteristics but also the set of possibilities, in such a manner
that the facts observed always relate to an equilibrium between the two kinds of factors of heredity and milieu and
not just to a single one of them.

9. Generally speaking, the organism constantly assimilates the milieu to its structure while, simultaneously, it
accommodates the structure to its milieu; the adaptation
can be defined as an equilibrium between such changes.

10. From the psychological viewpoint, this equilibration notion plays a considerable role, and its importance is perhaps due only to our ignorance of the limitations between the innate and the acquired. The merit of the gestalt theory, for example, is to show that perceptive forms can be explained by laws of equilibrium independent of the limits in question. The schemes of sensorimotor intelligence can be interpreted by a progressive equilibrium between assimilation and accommodation. We can continue this functional description at the preoperatory and operatory levels of thought itself.

11. From such a viewpoint, the specific result of thought would be to achieve a permanent equilibrium between assimilation of the universe to the subject and accommodation of the subject to the objects, whereas the organic or sensorimotor and chiefly perceptive forms know only constant equilibration displacements. In other words, the reversible play of anticipation and of mental reconstitution would achieve a form of equilibrium both mobile and stable in opposition to static and unstable configurations.

Structures

12. To commit oneself to studying equilibration forms means questioning the significance of structures. The problem therefore is to free the principal cognitive structures and to seek their relation with organic structures. Any birth leads to a structure and any structure is a form of terminal equilibrium including a birth. The opposite view

which sought to introduce phenomenology between birth and structure thus remains artificial.

13. There exist two extreme types of cognitive structures linked by numerous intermediary small chains: the perceptive gestalt of nonadditive and irreversible composition; and the operatory structures of intelligence of additive composition, founded on the two complementary forms of reversibility: inversion or negation and reciprocity (groupments, groups, and networks).

14. Gestalt structures can be found on the organic level. Even if we have not verified the consequences of the gestalt theory on the level of cerebral organization, a certain number of organic forms stem from the gestalt, (for example, during the course of early embryonic stages).

15. Psychologically, and notably on the ground of perceptive forms, the gestalt characteristics can doubtless be explained by a probabilist mode of composition. The field effects thus stem from a kind of sampling, the cause of distortions, and, in the case of good forms, these distortions would compensate to the *maximum*. Thus the nonadditive composition peculiar to gestalt would not be of a nature to confer to the *whole* a special power of emergence. If it is distinct from the sum of the parts (more but often less), this is merely because of the indetermination of the compositions.

16. Perception does not stem, moreover, exclusively from such mechanisms. One above another on several levels, it certainly begins by such field effects, but it is then structured by a perceptive activity of nature, among other

things propulsive and postural, and at a certain level this activity can itself be oriented by the operatory activity. When we attribute to perception a play of implications, of unconscious reasoning (Helmholtz), a "prolepsis" (von Weizsäcker), and so forth, it is therefore worthwhile to determine carefully to which types of connections or regulations these words respond when they correspond to something.[1]

17. The operatory structures which develop in the child from the age of seven to eleven or twelve (groupments of classifications and relations), then from there to the age of fourteen or fifteen (groups and networks of interpropositional operations), reveal the progressive reversibility of intelligence and thus appear more and more removed from known organic structures.

18. However, there remains the problem of knowing whether the vital processes are wholly submitted to irreversibility conforming to the second principle of thermodynamics (growth of entropy with the probabilist models which have been furnished) or whether, as Helmholtz, Guye, and many others believed, the living organism supposes a mechanism escaping from this principle and then converging with the characteristic reversibility of intelligence.

19. In the restricted field of nervous structures in par-

---

[1] We have shown with Lambercier that *prolepsis* was useless to explain square effects in circulation proposed by Auersperg and Buhrmester to prove Weizsäcker's thesis. Nevertheless there are certain facts of perceptive anticipation.

ticular, a set of present considerations speaks in favor of such convergence: McCulloch's and Pitts's applications of logical structures to the neuronic connections, Rashevsky's and Rapoport's efforts to form a mathematical neurology implying the same structures (the law of all-or-nothing supposes an isomorphic arithmetic module 2 to Boole's algebra), cybernetic work, especially the role attributed to feedback as semireversible regulation, and the use of group and network structures to characterize the equilibration phases and the arrival at the equilibrium terminal.

20. Although we manage to glimpse such correspondences between the operatory structures of thought and certain physiological structures, nevertheless this does not show the innate characteristic of the first. What furnishes a hereditary structure (and this can be applied to every problem raised by the maturation of the nervous system) is the list of possibilities and impossibilities of a given level. It is therefore a question of a system of "virtual work" whose compensation defines a state of equilibrium. But if a correspondence is possible between a form of organic equilibrium and a form of mental equilibrium, the problem entirely remains of the conditions of the actualization of operations and the progressive construction of the system of effective or virtual operations.

Parallelism or psychophysiological isomorphism

21. Having posed the problem of psychophysiological parallelism, several authors buried themselves artificially in

the following alternative: Either a parallelism exists between the states of consciousness and certain physiological states, and in that case consciousness is only a reflection without activity proper, or else consciousness acts, and in this case it intervenes causally in the organic mechanisms (interaction).

22. Such an alternative stems from the fact that partisans or adversaries of one or the other of the two solutions agree in reasoning only according to certain categories (substance, energy, work, causality, and so forth). Rightly refusing the attribution of these categories to consciousness, the parallelists then believe in removing from it, or are accused of so doing, any efficiency, whereas the interactionists restore to it such modes of existence or of activity at the cost of insoluble conflicts with biology.

23. Supposing, on the contrary, that consciousness exclusively constitutes a system of implications (in the large sense) between significations, a system whose superior forms consist in logical necessities or in moral obligations (implications among values, juristic imputation in the sense of Kelsen's normativism, and so forth), and whose inchoative forms remain at the state of more or less structurized relations between signals and indications. In this case, any substantiality, energy, causality, and so forth, will remain peculiar to the material or organic connections, nevertheless consciousness will present an irreplaceable or specific originality. Consciousness is the source of logic and of mathematics, and it will constitute the complementary aspect indispensable to the causal series.

24. Thus, without falling into the preceding alternative, we can conceive not an end-to-end parallelism but a structural isomorphism between the system of conscious implications and certain systems of organic causality.

25. This isomorphism of conscious implication and of organic causality can be conceived as a special case of correspondences between deduction and material reality which characterize the whole circle of sciences. Let us suppose the logico-mathematical structures placed in sufficient isomorphism with organic structures, then the latter explained causally in an efficient manner by a "generalized" physico-chemistry, to quote Guye, until the biological fact is encompassed. This physico-chemistry itself could not help becoming mathematical and deductive, thus based, as a point of departure, on its point of arrival. . . . It is in the perspective of such a circle or, if we prefer, of such a constantly increasing spiral that it is probably fitting to situate the problems of relation between life and thought.

References

CHAPTER 1. Time and the Intellectual Development of the Child. *La vie et le temps. Rencontres internationales de Genève.* Neuchâtel, La Baconnière, 1962.

CHAPTER 2. Affective Unconscious and Cognitive Unconscious. *Raison présente*, No. 19. Paris, Editions rationalistes, 1971.

CHAPTER 3. The Stages of Intellectual Development in the Child and Adolescent. *Le problème des stades en psychologie de l'enfant. Symposium de l'Association psychologique scientifique de langue française.* Paris, Presses Universitaires de France, 1956.

CHAPTER 4. Child Praxis. *Revue neurologique,* No. 102. Paris, Masson, 1960.

CHAPTER 5. Perception, Learning, and Empiricism. *Dialectica,* No. 13. Neuchâtel, Editions du Griffon, 1959.

CHAPTER 6. Language and Intellectual Operations. *Problèmes de psycho-linguistique. Symposium de l'Association de psychologie scientifique de langue française.* Paris, Presses Universitaires de France, 1954.

CHAPTER 7. What Remains of the Gestalt Theory. *Revue suisse de psychologie,* No. 13, 1954.

CHAPTER 8. Necessity and Significance of Comparative Research in Genetic Psychology. *Journal international de psychologie,* Vol. 1, No. 1. Paris, Dunod, 1966.

CHAPTER 9. Life and Thought. *La vie, la pensée. Actes du 7ᵉ Congrès des Sociétés de philosophie de langue française, Grenoble, September 12–16, 1954.* Paris, Presses Universitaires de France, 1954.

Index